Praise for

The Truth About Animals

"[Cooke's] pace is quick, her touch is light, and through her wealth of research we can reach new heights of wonder."
—*New York Times Book Review*

"Funny and fascinating, this book will change the way you see wildlife."
—*Bustle*

"A surefire summer winner . . . even Cooke's simple facts are funny."
—*New York Times Book Review*

"Cooke, founder of the Sloth Appreciation Society, raises the profile of many poorly understood animals, revealing surprising, and often hilarious, truths that are much better than the fictions."
—*Scientific American*

"[A] deeply researched, sassily written history of 'the biggest misconceptions, mistakes and myths we've concocted about the animal kingdom,' spread by figures from Aristotle to Walt Disney."
—*Nature*

"Endlessly fascinating."
—Bill Bryson, author of *A Walk in the Woods*,
A Short History of Nearly Everything, and *In a Sunburned Country*

"Lucy Cooke's modern bestiary is as well-informed as you'd expect from an Oxford zoologist. It's also downright funny."
—Richard Dawkins, author of *The God Delusion*,
The Selfish Gene, and *The Ancestor's Tale*,
and emeritus fellow of New College, Oxford

"Cooke's extensive travels, research, and delightful sense of humor make *The Truth about Animals* a fascinating modern bestiary."

—*Seattle Times*

"In the end, the history of zoology reveals as much about our human foibles as about the animals we study. And this book will leave readers more enlightened about both."   —*Science News*

"Lucy Cooke's new book makes *Playboy* seem as tedious and tame as a phone book. . . . In her delightful reading of natural history, [Cooke] is both scientist and standup comic. . . . Trained as an academic, Cooke's writing style is anything but—rather, it's bawdy, irreverent, guiltless, sometimes locker-room-ish and comedic. She makes you feel she's having fun as she pounds out the words. . . . *The Truth About Animals* is a great read and fascinating fun."   —*Winnipeg Free Press*

"Lucy Cooke takes equal delight in natural oddities and in people's long struggles to understand them. Part history, part biology, and wholly entertaining, *The Truth About Animals* is a quirky and edifying romp."

—Thor Hanson, author of *Buzz, Feathers,* and *The Triumph of Seeds*

"As surprising as it is diverse. Consummate natural history writing: illuminating, remarkable—and very, very funny."

—Alice Roberts, author of *The Complete Human Body* and
The Incredible Unlikeliness of Being

"[An] intriguing and amusing survey of some unusual facets of animal behavior . . . Cooke puts scientific errors, some of them hilarious, into historical context."   —*Booklist*

"[A] lighthearted but scientifically rigorous exploration. . . . A pleasure for the budding naturalist in the family—or fans of Gerald Durrell and other animals."   —*Kirkus Reviews*

"Readers keen on animals and natural history in general should find Cooke's discussion fascinating and educational."

—*Publishers Weekly*

"Lucy Cooke's [*The Truth About Animals*] was a joy from beginning to end. Who could resist a writer who argues that penguins have been pulling the wool over our eyes for years, and that, far from being cute and gregarious, they are actually pathologically unpleasant necrophiliacs?"

—*The Guardian* (UK)

"Each of Cooke's thirteen breezy yet fact-stuffed chapters traces the origins of a long-standing myth about a species or class of animal."

—*Times Literary Supplement* (UK)

"The eclectic stories come thick and fast, with an equally varied human cast dedicated to uncovering the truth, scientifically or otherwise. Cooke illuminates and mickey-takes in equal measure, and the truth as she tells it is not only unexpected but often bizarre, bawdy and very, very funny."

—*BBC Wildlife* (UK)

"The rising star of natural history...is she the new David Attenborough?"

—*The Times* (UK)

"Lucy Cooke's fascinating book is full of mind-boggling stuff. Cooke takes much pleasure in throwing in all manner of amazing facts."

—*Reader's Digest* (UK)

"A riot of facts...Cooke scores a series of goals with style and panache."

—*The Times* (UK)

"Beautifully written, meticulously researched, with the science often couched in outrageous asides, this is a splendid read. In fact, I cannot remember when I last enjoyed a nonfiction work so much."

—*Daily Express* (UK)

THE TRUTH
ABOUT ANIMALS

ALSO BY LUCY COOKE

A Little Book of Sloth

THE TRUTH
ABOUT ANIMALS

Stoned Sloths, Lovelorn Hippos,
and Other Tales from
the Wild Side of Wildlife

LUCY COOKE

BASIC BOOKS
New York

Basic Books
Hachette Book Group
1290 Avenue of the Americas, New York, NY 10104
www.basicbooks.com

Printed in the United States of America
Originally published as *The Unexpected Truth About Animals* in hardcover and ebook by Transworld Publishers in November 2017 in the United Kingdom

First Trade Paperback Edition: April 2019
Originally published in the US in hardcover and ebook by Basic Books in April 2018
Published by Basic Books, an imprint of Perseus Books, LLC,
a subsidiary of Hachette Book Group, Inc.

The Basic Books name and logo is a trademark of the Hachette Book Group.

The Hachette Speakers Bureau provides a wide range of authors for speaking events. To find out more, go to www.hachettespeakersbureau.com or call (866) 376-6591.

The publisher is not responsible for websites (or their content) that are not owned by the publisher.

Additional copyright/credits information is on page 271.

Editorial production by Christine Marra, Marrathon Production Services.
www.marrathoneditorial.org

Book design by Jane Raese
Set in 12-point Dante

Library of Congress Control Number: 2018966585
ISBN 978-0-465-09464-6 (hardcover)
ISBN 978-0-465-09465-3 (ebook)
ISBN 978-1-5416-7408-0 (paperback)

LSC-C

10 9 8 7 6 5 4 3 2 1

TO THE MEMORY OF MY DAD,

who opened my eyes to the wonders

of the natural world

Contents

THE TRUTH
ABOUT ANIMALS

"How can sloths exist when they're such losers?"

As a zoologist and founder of the Sloth Appreciation Society I get asked this question *a lot*. Sometimes "losers" is further defined—"lazy," "stupid" and "slow" being perennial favorites. And sometimes the query is paired with the rider—"I thought evolution was all about survival of the *fittest*"—delivered with an air of bemusement or, worse, a whiff of superior species smugness.

Sloths are, in fact, one of natural selection's quirkiest creations, and fabulously successful to boot. Skulking about the treetops barely quicker than a snail, and being covered in algae, infested with insects and defecating just once a week might not be your idea of aspirational living, but then you're not trying to survive in the highly competitive jungles of Central and South America—something the sloth is very good at.

The reputation of the sloth was sufficiently besmirched that I felt compelled to found the Sloth Appreciation Society. (Our motto: "Being fast is overrated.") I gave a talk on the unexpected truth about this much maligned creature to festivals and schools. This book grew out of those talks and the need to set the record straight—not just for the sloth, but for other animals as well.

When seeking to understand animals, context is key. We have a habit of viewing the animal kingdom through the prism of our own,

I love sloths. What's not to like about an animal born with a fixed grin on its face and the desire to hug.

rather narrow, existence. The sloth's arboreal lifestyle is sufficiently extraterrestrial to make it one of the world's most misunderstood creatures, but it is by no means alone in this category. Life takes a glorious myriad of alien forms, and even the simplest require complex understanding.

Evolution has played some splendid practical jokes by fashioning implausible creatures with an absence of logic and precious few clues to explain itself. Mammals like the bat that want to be birds. Birds like the penguin that want to be fish. And fish like the eel, whose enigmatic life cycle sparked a two-thousand-year search for its missing gonads. Animals do not give up their secrets easily.

CONSIDER THE OSTRICH. In February 1681, the brilliant British polymath Sir Thomas Browne wrote a letter to his son Edward, a physician at the royal court, requesting a rather unusual favor. Edward had come into possession of an ostrich, one of a flock donated to King Charles II by the king of Morocco. Sir Thomas, a keen naturalist, was fascinated by this big foreign bird and eager that his son send him news of its habits. Does the bird sleep with its head under its wing? Is it vigilant like a goose? Does it delight in sorrel but recoil from bay leaves? And does it eat iron? This final query, he suggested helpfully to his son, might be best uncovered by wrapping the metal first in pastry since "perhaps it will not take it up alone."

Browne wanted to test an ancient myth that ostriches were capable of digesting absolutely anything, even iron. According to the medieval German scholar Sebastian Müenster, the ostrich's taste for the strong stuff was such that the bird's dinner "consists of a church-door key and a horse shoe." As ostriches were bestowed on the courts of Europe by the emirs and explorers of Africa, generations of enthusiastic natural philosophers encouraged the foreign fowl to consume scissors, nails and a glut of other ironworks.

On the surface this experimentation appears to be lunacy, yet dig a little deeper and there is a (scientific) method to the madness. Ostriches can't digest iron, but they have been observed swallowing large, sharp stones. Why? The world's biggest bird has evolved into a rather unusual grazing animal, whose usual diet of grasses and shrubs is tough to digest. And unlike their fellow plant munchers from the African plains, the giraffe and antelope, ostriches lack a ruminating stomach. They don't even have teeth. Instead, they must tear the fibrous grasses from the ground with their beak and swallow them whole. They employ the quarry of jagged rocks in their muscular gizzard to do the job of grinding down this stringy dinner into more digestible pieces. They can clunk around the savannah with up to a kilogram of stones in their stomach. (Scientists fancy this up and call them *gastroliths*.)

Again, understanding the ostrich is about context. But so too we must understand the context of the scientists who have been prodding and poking for the truth about animals for centuries. As such, Browne is just one of a great cast of idiosyncratic obsessives. There's the seventeenth-century physician who tried to spontaneously generate toads by placing a duck on a dung heap (an old recipe for creating life). There's also an Italian Catholic priest who wielded a mean pair of scissors in the name of science, whether tailoring tiny, bespoke underpants for his animal subjects or removing their ears.

Scientists in more recent times have also chosen to pursue bizarre, and often misguided, methods in their search for truth—like Ronald K. Siegel, the twentieth-century American psychopharmacologist whose curiosity compelled him to get a herd of elephants very drunk indeed, with suitably demented results. Every century has its eccentric animal experimenters, and there will no doubt be many more to come. We humans may have split the atom, conquered the moon and tracked down the Higgs boson, but when it comes to understanding animals we still have a long way to go.

I'm fascinated by the mistakes we've made along the way and the myths we've created to fill in the gaps in our understanding. They reveal much about the mechanics of discovery and the people doing the discovering.

I've found that taking a dissecting knife to our greatest animal myths often exposes a charming logic, transporting us to times of wondrous naivety, when little was known and anything was possible. Why on earth wouldn't birds migrate to the moon, hyenas switch sex with the season and eels spontaneously generate out of mud? Especially when the truth is no less incredible.

I STUDIED ZOOLOGY in the early 1990s at New College, Oxford under the great evolutionary biologist Dr. Richard Dawkins and was taught a method of thinking about the world based upon the genetic relationships among species—how their degree of relatedness influences their behavior. Some of what I was taught has already been surpassed by recent advancements, which show that the *way* in which a genome is read at a cellular level is at least as important as its content (which is how we can share 70 percent of our DNA with an acorn worm and yet be so much more fun at a dinner party). Each generation—mine included—thinks it knows more about animals than its predecessors, yet we're still often wrong. Much of zoology is little more than educated guesswork.

With modern technology we are getting better at our guessing. As a producer and presenter of natural history documentaries, I've traveled the world and gained privileged access to some of the most dedicated scientists battling for truth at the front line of discovery. I've met an animal IQ tester in the Maasai Mara, a peddler of panda porn in China, the English inventor behind a sloth "bum-o-meter" and the Scottish author of the world's first chimpanzee dictionary. I have chased after drunk moose, nibbled beaver "testicles," savored amphibian aphrodisiacs, jumped off a cliff to fly with vultures and attempted to speak a few words of hippo. These experiences have opened my eyes to many surprising realities about animals and the progress of animal science. This book is my effort to share these revelations with you, and to rebrand the animal kingdom with fact, not fiction. I have gathered together the biggest misconceptions, mistakes and myths we've concocted about the animal kingdom, whether the purveyor was the great philosopher Aristotle or the Hollywood descendants of Walt Disney, and create my very own menagerie of the misunderstood. Open your mind to their incredible tales and get ready to discover the truth about animals.

CHAPTER I

Genus *Anguilla*

*There is no animal concerning whose origin and existence there
is such a number of false beliefs and ridiculous fables.*
—Leopold Jacoby, "The Eel Question," 1879

Aristotle was obsessed with eel genitals. He may have been the first
true scientist and the father of zoology, and he certainly made acute
scientific observations about hundreds of creatures, but he was out-
foxed by eels. No matter how many the great Greek thinker sliced
open, he could find no trace of their sex. Every other fish he'd examined on his island laboratory of Lesbos had easily detectable (and often
quite delectable) eggs and conspicuous, albeit internal, testicles. But
the eel appeared to be entirely sexless.

When Aristotle came to writing about this slimy enigma in his pioneering animal almanac in the fourth century BC, he was forced to
conclude that the eel "proceeds neither from pair, nor from an egg"
but was instead born of the "earth's guts"—that is, it spontaneously
emerged from mud; he thought the worm casts we see in wet sand
were embryonic eels boiling out of the ground. He folded the origins
of the eel into his theory of spontaneous generation, which he applied liberally to an eclectic collection of critters—from flies to frogs—
whose proliferation seemed inexplicable.

According to Aristotle's *Historia Animalium*, certain lower animals "are not produced from animals at all, but arise spontaneously: some are produced out of the dew which falls on foliage . . . others are produced in putrefying mud and dung, others in wood, green or dry, others in excrement whether voided or still within the living animal."

Aristotle's theory is bizarre, but no more so than the truth. These slippery characters keep their secrets especially well hidden. The so-called common freshwater eel, *Anguilla anguilla*, starts its life as an egg suspended in the depths of an underwater forest in the Sargasso Sea, the deepest, saltiest slice of the North Atlantic. As a wisp of life no bigger than a grain of rice, it embarks on an odyssey to the rivers of Europe lasting up to three years, during which it undergoes a transformation as radical as a mouse turning into a moose. It then spends decades living in the mud and fattening itself up only so it can repeat the grueling nearly four-thousand-mile journey back to its obscure oceanic womb, where it spawns in the shadowy recesses of the continental shelf and then dies.

THE FACT THAT the eel doesn't become sexually mature until the very tail end of its long, peculiar life, after its fourth, and final, metamorphosis, has helped to obscure its origins and bestowed it with a mythical status. Over the centuries, unraveling the mystery has pitted nations against each other, driven man to the remotest reaches of the seas and tormented an eclectic cast of eel enthusiasts including obsessive gonad hunters, gun-toting fishermen, the world's most famous psychoanalyst—and me.

As a child, I too was rather obsessed with eels. When I was about seven, my father sunk an old Victorian bathtub in the garden, and converting this sterile vessel designed for human ablution into the perfect pond ecosystem became my principle pastime. Every Sunday my dad would accompany me to the ditches of Romney Marsh, a wetland in the South of England close to my hometown of Rye. There I would spend happy hours trawling for any form of life with an improvised underwater animal trap he'd fashioned for me out of a pair of old net curtains. At the end of the day we'd return triumphant, heady with the zeal of Victorian explorers, our underwater booty sloshing around in the back of his aged mini pickup, ready to be identified and introduced to my watery kingdom. The animals came two by two: marsh frogs, smooth newts, sticklebacks, whirligig beetles and pond skaters all joined the party in my bath. But no eels. My trusty net collected them, but attempting to transfer their slimy bodies into the bucket was like trying to hold on to water. Every time I grabbed one, it'd escape, slithering off to safety overland—more like a snake than a fish out of water. Catching one became my holy grail.

What I didn't know was that, if I *had* succeeded in my mission, the eels would have brought an end to my pleasant pond party by eating all the other guests. Eels spend the freshwater phase of their life like extreme prizefighters bulking themselves up for a championship bout in preparation for the long swim back to the Sargasso to breed. To achieve this, they will eat anything that moves—including each other.

Their rapacious appetite was exposed in a gruesome experiment conducted by a pair of French scientists in Paris in the late 1930s. The researchers placed a thousand elvers—young eels, about three inches long—in a tank of water. The fish were fed daily, but even so, a year later there were only seventy-one eels left, now three times as long. Another three months on, after what a local journalist reported as "daily scenes of cannibalism," one champion was left: a female measuring a foot in length. She lived four more years all on her own, until she was accidentally bumped off by the Nazis, who inadvertently cut off her supply of worms during their occupation of Paris.

This horror story would have shocked past generations of naturalists who believed the eel to be a benign vegetarian with a particular weakness for peas—they were said to leave their watery world and seek out their favorite juicy legumes on land. Such accounts came courtesy of the thirteenth-century Dominican monk Albertus Magnus, who, in his book *De Animalibus*, noted: "The eel also comes out of the water in the night time where he can find pease, beans, and lentils."

The eel's hippie diet was still in currency in 1893, when *The History of Scandinavian Fishes* embellished the monk's "observations" with delicious sound effects. The eels that invaded the Countess Hamilton's

estate gobbled up her legumes "with a smacking sound, like that made
by sucking pigs when they are feeding." Though perhaps lacking the
requisite manners, the dowager's eels were a suitably discerning school
that "only [ate] the outer soft and juicy skin surrounding the pods" and
discarded the rest. While it's true that eels can survive for a remarkable forty-eight hours out of water, thanks to their slimy, breathable
skin—an adaptation that allows them to pond-hop in search of water
in times of drought—reports of their lip-smacking, pea-stealing antics
were quite delusional.

While the eel's greedy freshwater years lead to an impressive increase in size, the great Roman naturalist Pliny the Elder's assertion,
made in his epic tome, *Naturalis Historia*, that eels from the river Gan-
ges grew to be "thirty feet long" was a cocky overstatement even in
this well-worn genre of lies.

Izaak Walton, author of a seventeenth-century fishing bible *The
Compleat Angler*, showed a bit more restraint when describing an eel
caught in the river Nene of east England that he claimed "was a yard
and three quarters long." Walton was keen to fend off any doubters,
adding, perhaps a little too quickly: "If you will not believe me, then
go and see it at one of the coffee-houses in King street Westminster"
(where it was no doubt happily sipping cappuccinos and regaling cus-
tomers with stories of its youthful adventures at sea).

More measured measurements come from Dr. Jorgen Nielsen
of the Zoological Museum in Copenhagen, who had examined the
corpse of an eel from a rural pond in Denmark. He told Tom Fort,
author of *The Book of Eels*, that the prizewinning specimen punched in
at just over four feet. Unfortunately, the slippery monster had met an
untimely death when the pond's owner caught it menacing his beloved
ornamental waterfowl and came at it with his shovel.

The eels I caught were not much more than the length and thickness of a pencil. They were no doubt nearer the start of their freshwater life, which can last anything from six to thirty years. Some eels
have been known to live much longer. A Swedish specimen nicknamed
Putte, caught as an elver near Helsingborg in 1863 and kept in a local
aquarium, passed away aged eighty-eight. Her death was mourned

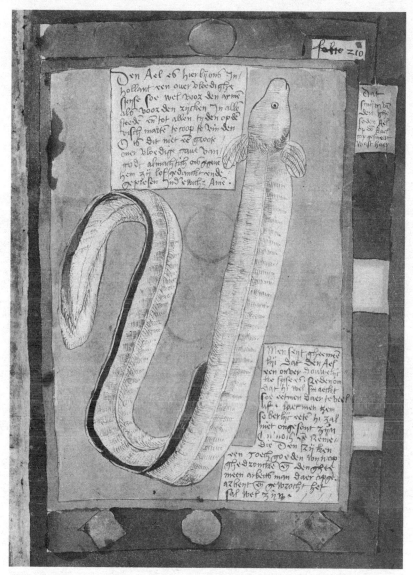

The eel in Adriaen Coenen's almanac of fish (1577) is a veritable monster that measures a whopping "40 feet" long (having grown a further ten feet since being described by Roman naturalist Pliny the Elder).

with extensive media coverage, her record-breaking age having afforded her the kind of celebrity status not normally available to a long, slimy fish.

Such aged eels have invariably been denied their natural instinct to migrate back to the sea by being kept captive, often as pets. An eel may seem an unconventional choice for an animal companion—not a lot of fun to be had in the snuggling department—but the Roman orator Quintus Hortensius was said to weep at the death of his, "which he kept long, and loved exceedingly."

THE EEL'S FRESHWATER EXISTENCE may be long and gluttonous, but it is just one of the fish's many lives, albeit the only one obvious to me and countless other naturalists over the centuries. It gives no clues about the rest of its life cycle—its birth, reproduction and death, which are shrouded by the sea—spawning an intense international quest, lasting some two thousand years, to locate the eel's missing gonads.

Aristotle was one of the first to be stymied by the genesis of this apparently sexless fish, but he was hardly alone. Several hundred years later Pliny the Elder trialed his own imaginative ideas about eel propagation by proposing that they reproduced by rubbing themselves against rocks: "the scrapings come to life." Hoping to have the last word on the matter, the Roman naturalist concluded with an authoritative flourish: "This is the only way they breed." Alas, Pliny's asexual friction was nothing but fiction.

Fantastic rumors of the eel's reproduction bred like rabbits over the ensuing centuries. Eels were said to emerge from the gills of other fishes, from sweet morning dew (but only during certain months) and from enigmatic "electrical disturbances." One mysterious "reverend bishop" told the Royal Society he had seen young eels that had been born on the thatching of a roof. The eggs, he claimed, had been stuck to the thatching reeds and incubated by the heat of the sun.

Not all ecclesiastical naturalists were so open-minded about such fishy tales. In his *History of the Worthies of England*, Thomas Fuller poured scorn on the belief widely held in the fenlands of Cambridgeshire that the illicit wives and bastard children of priests were saved from damnation by taking the form of an eel. "No doubt the first founder of so damnable an untruth hath long since received his reward," he said.

The scientific geniuses of the Enlightenment swept aside such fanciful fables with significantly less silly—but no more accurate—theories of their own. In 1692, Antonie van Leeuwenhoek, the Dutch pioneer of microscopic worlds who discovered both bacteria and blood cells, erred towards credibility with his hypothesis that eels, like mammals, were *viviparous*, that is, their eggs were fertilized internally, with the females giving birth to live young. Van Leeuwenhoek at least embraced contemporary scientific method by basing his supposition on actual observations. He had stared down his magnifying lens and watched what appeared to be baby eels in what he assumed was the fish's uterus. Unfortunately, these were actually parasitic worms squatting in the eel's bladder, and had in fact been observed and dismissed as such by Aristotle nearly two thousand years before.

The eighteenth-century Swedish botanist and zoologist Carl Linnaeus also said that eels were viviparous, claiming to have seen what he believed to be baby eels inside an adult female. Surely no one would argue with the great father of taxonomy—a man so pedantic he even latinized his own name. But the awkward truth was that Linnaeus hadn't actually dissected an eel but an eel imposter, a similar-looking beast now known as the eelpout—an unusually viviparous yet altogether unrelated species of fish. Which is not to say his critics had their facts any straighter. One authority who reviewed Linnaeus's work took him to task for a case of mistaken identity—but, influenced by Aristotle, proclaimed the young eels discovered by the Swede to be parasitic worms, sending the live-birth doctrine into a vortex of inaccuracy and confusion.

Into this lofty academic fray stepped a plucky outsider. In 1862, a Scotsman by the name of David Cairncross announced to the world that he, a humble factory engineer from Dundee, had finally solved the riddle of the eel. "The reader may at once be informed that . . . the progenitor of the silver eel is a small beetle." His enthusiastic, if scientifically baffling, theory—the product of sixty years of ongoing experiments, by his own reckoning—took the form of a short book, *The Origin of the Silver Eel.*

Cairncross began his treatise with an apology for his lack of any interest in learning the rules and norms of contemporary science. "I could not be expected to be acquainted with the names and terms used by naturalists in their classifications of the different animals, my knowledge of such books being limited," he noted. His unconventional but highly convenient solution was to "employ names and terms of my own," which included a reimagining of animal classification into three nonsensical classes that would have made the great Linnaeus turn in his grave.

Cairncross's story begins at the tender age of ten, when the curious lad observed a number of "hair eels" (his term) in an open drain. "Where can they come from?" he wondered. A friend related to him a common folk belief that young eels "fall from the tails of the horses while drinking; and the water brings them to life." The young Cairncross scoffed at this highly improbable explanation, before conjuring up his own, equally implausible idea inspired by a number of dead beetles languishing in the bottom of the same drain. Maybe the two animals were connected?

This drain-based miniature drama haunted the Scot for two decades. Then one summer, the adult Cairncross spotted a familiar-looking beetle in his garden in Dundee. He watched it intently, attempting to read its thoughts as it walked with determination towards a puddle and plunged in. The beetle, he reported, then "looks about a bit" before exiting its bath "in a very troubled state." How Cairncross arrived at his diagnosis of the beetle's mental state is not known. But the book's only illustration provides the reader with valuable assistance in comprehending the insect's next extraordinary move: entitled "the beetle in the act of parturition," it shows Cairncross's unlikely hero lying on its back with what appears to be a pair of lassoes emerging from its backside. The beetle, according to the Scotsman, was giving birth to two fish.

This was the eureka moment for Cairncross. He now dedicated himself to furthering his investigation by cutting open beetles, removing

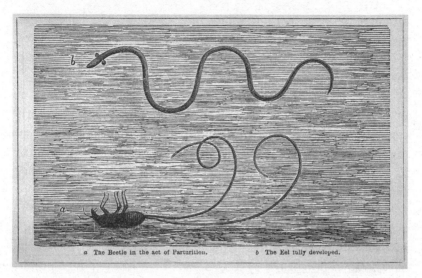

a The Beetle in the act of Parturition.   *b* The Eel fully developed.

Just in case you had any problems visualizing how a beetle might give birth to a pair of eels, *The Origin of the Silver Eel* included this charming illustration in an attempt to authenticate the author's wild claims. Nice try, Cairncross, but I'm still not convinced.

"hair eels" and keeping them alive for various, albeit rather limited, periods of time. He freely admitted that his theory "may seem strange" but reassured himself by looking at the behavior of "members of the vegetable kingdom." If one species of tree can be grafted onto another, "Could not therefore the Great Creating Gardener graft a foreign nature on that of an insect?"

All manner of Frankenstein animals have been conceived in modern labs: human ears have been grafted onto mice and glow-in-the-dark fish have been created with a judicious sprinkling of jellyfish genes. But the "Great Creating Gardener" had no hand in planting them.

Had Cairncross posed his question to the scholarly community, they would have declared that his "hair eels" were yet another case of pesky parasitic worms and not the fish in an early stage of development. But the factory manager knew not of peer review. He presented his exceptional findings not to the Royal Society for serious scrutiny, but instead to a pair of farmers he bumped into one day who were perplexed by the quantities of silver eel in a ditch on their land. So he explained his theory that this profusion of eels had emerged from a beetle's bum. "They believed me," he announced with pride, "and rejoiced in the solution of the mystery."

BEAVERING AWAY IN intellectual isolation for sixty-odd years, Cairncross was unaware of the radical progress being made in the hunt for the fish's gonads. Far from Dundee, Europe's scientific intelligentsia were gripped by "the eel question," and they were about to reach a climax—of sorts.

Leading the charge were the Italians, who embraced the quest to locate the eel's missing sex organs as an unlikely source of civic pride for their troubled nation. Early modern Italy was a divided mess, with huge swathes dominated by a succession of warring foreign powers and its many independent states far from unified.

There was one thing, however, that united these fractious states, and that was a love of eel. The Italians had cultivated a long-standing relationship with the slippery fish that mostly involved eating them in vast quantities. Eel is an unusually fatty fish—an evolutionary adaptation to fuel its onerous nearly four-thousand-mile odyssey back to its mating grounds deep in the Sargasso Sea. Unfortunately for the eel, this high lipid content also makes it especially tasty. The Roman epicure Marcus Gavius Apicius, author of what is arguably the world's first cookbook, wrote in the first century AD that six thousand eels were served at feasts celebrating the victory of Julius Caesar. He recommended that the "eel will be made more palatable" by serving with a sauce of "dry mint, rue berries, hard yolks, pepper, lovage, mead, vinegar broth and oil." (In England we still like to eat them simply boiled and jellied, surely one of the greatest crimes against gastronomy the British have committed in a long and illustrious history of murdering food.) Yet, despite such crude recipes, eels were long associated with great feasts and gluttony. Leonardo da Vinci painted the disciples enjoying eels at the Last Supper, and a surfeit of the slippery fish was blamed for the death of the infamous gourmand Pope Martin IV.

The finest-tasting eels were said to come from Comacchio and the surrounding vast gray wetlands of the mighty river Po delta. They were home to Europe's greatest eel fisheries, hauling in three hundred tons of eel a night at the height of the season, and the source of some of the greatest pronouncements and controversies surrounding the eel's sex. These began in 1707, when a local surgeon noticed an unusually plump eel among the many thousands being caught. When the doctor sliced it open, he saw what looked to him like an ovary stuffed with ripe eggs. The pregnant fish was dispatched to his friend, the esteemed naturalist Antonio Vallisneri, who hastily proclaimed the centuries-long search for the eel's private parts was finally over. The learned professor had already lent his name to the formal classification of a water plant known colloquially as eelgrass, but the genitals of the female eel would not be a further namesake. Upon closer scrutiny, the discovery was written off as a diseased and distended swim bladder.

Vallisneri's flirtation with victory inspired the Italian scientific establishment, who now considered it "a matter of extreme importance to find the true ovaries of the eel." These were turbulent times for the inchoate nation, which, as an unstable hodgepodge of warring independent states and squatted by European superpowers like Austria, was struggling with its identity. And while many Italians hung their nationalistic hopes on revolution, this small band of intellectuals dreamed instead of empowering their countrymen by laying claim to the delectable eel's elusive gonads.

The professors devised a plan. Thousands of eels were captured daily around Comacchio; all they need do was offer an enticing reward to the first fisherman who could provide them with a specimen complete with roe. In Germany, a similar plan backfired when the naturalist who had hatched it received so much eel offal in the post that he was moved to "cry and beg for mercy." The Italian scheme, however, quickly yielded positive results—or so it seemed. The celebrations were cut short when it was discovered the wily fisherman had simply filled his eel with the eggs of another fish. This humiliating blow dampened the Italians' zeal for eels for some fifty years.

Then in 1777, a fresh, fat, slimy suspect flopped up on the shores of Comacchio. It was immediately examined by the anatomist Carlo Mondini, professor at the nearby University of Bologna, who made an ingenious realization: the frilled ribbons inside the eel's abdomen were not fringes of fatty tissue, as had been previously supposed, but the female eel's evasive ovaries.

Italy's scientists rejoiced, perhaps, again, a little prematurely. After all, the eel's testicles were still missing, along with any clear notion of how this enigmatic fish reproduced. Thus, the mission to complete the eel's genital jigsaw puzzle fell to a rather unlikely candidate: an ambitious young medical student who would later find fame locating the seat of desire, in humans if not in eels. The student's name: Sigmund Freud.

As A NINETEEN-YEAR-OLD STUDENT at the University of Vienna, the founding father of psychoanalysis took up his very first research job in 1876 at a zoological field station in Trieste, on Italy's Adriatic coast. His task was to investigate the claim of a Polish professor named Szymon Syrski, who had said he had discovered the eel's testes.

The only way to determine gender was to slice the fish open, "seeing that eels keep no diaries," Freud opined sardonically in a letter to a friend. For weeks he did just that, every day, from eight in the morning until five in the afternoon, in a hot, smelly laboratory.

Four weeks and four hundred disemboweled eels later, Freud gave up. "I have been tormenting myself and the eels, but in vain, all the eels which I cut open are of the fairer sex," he lamented in a letter littered with doodles of eels wearing thin mocking smiles. Freud's resulting paper, "Observations on the Form and the Fine Structure of Looped Organs of the Eel, Organs Considered as Testes," was his first published work. Although he suspected Syrski was right, he could neither confirm nor deny the Pole's claims. "But since he apparently doesn't know what a microscope is," grumbled Freud in his letter, "he failed to provide an accurate description of them."

It's anyone's guess how much those long days spent slicing open phallocentric fish in a fruitless search for their sex influenced Freud's

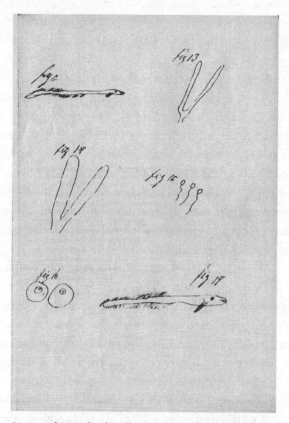

Sigmund Freud's doodles in a letter to a friend
are a window into his mind during his futile hunt
for the testicles of the eel. Here are the enigmatic
eels that tortured him so, along with squiggles
representing their elusive sperm and eggs (which
a psychoanalyst might say look suspiciously like a
pair of breasts).

later theories surrounding the phallic-envy stage of human psycho-sexual development. In any case, he went on to probe less slippery subjects, like the human psyche, with significantly more success.

TWO DECADES LATER, a lone male eel finally exposed his private parts. The fortuitous young biologist to make the eel's acquaintance was another Italian, Giovanni Battista Grassi, who apprehended the fish—his sex organs swollen with sperm—swimming off the coast of Sicily. Grassi had already produced some seminal work on the anatomy of termites, and had named a new species of spider after his wife. But he was on something of a roll when it came to eels. Not only had he won the international eel testicle tournament for Italy, but the year before he had made an equally pivotal discovery by identifying a key stage in the eel's inscrutable life cycle.

As early as the 1850s, tiny, transparent fish the shape and thickness of a willow leaf, with bulbous black eyes and gruesome buck teeth, had been documented washing up in great numbers on the shores of Italy. These minuscule monsters were classified as *Leptocephalus brevirostris*, which translates to "thin-head short-nosed," and quickly dismissed as just another of the all too numerous, utterly nondescript marine creatures that inhabit the murky depths. Grassi was fascinated by these slivers of life. Suspecting they might be the larval stage of another fish, he pulled a rather cunning trick. He counted up their embryonic vertebrae—which averaged 115—and looked for a match in an adult species. He found it in the European freshwater eel. Grassi's piece of piscine detective work had yielded a truly momentous revelation—the identification of the missing link in the eel's mysterious life cycle.

Several learned minds had already proposed that the freshwater eel must breed far out at sea. It was an unconventional idea—the reverse direction of all other long-distance migrating fish, like salmon, which spread their lives across fresh- and saltwater habitats. But why else would eels swim downriver every autumn in such great numbers and with such determination, with miniature versions of themselves making the reverse journey upriver every spring? Still, there was no evidence supporting this logical hypothesis. No baby eels had ever been found at sea. Now Grassi had not only located the missing larval stage but he had also fingered the eel as a world-class shape-shifter.

Grassi set himself up in an aquarium to observe the miraculous metamorphosis with his own eyes. He was clever to do so, as no one would probably have believed him otherwise. Over the course of several weeks, the leaflike wisp of life began to thicken at each end, forming an indisputably cylindrical, eel-shaped creature. Its body length shrunk by almost a third, its jagged teeth dissolved and, for obscure alimentary reasons, its anus migrated. After a few days, a perfectly transparent, bug-eyed noodle known as a glass eel was swimming around the tank. Giddy with discovery, Grassi proclaimed the Strait of Messina, off the coast of Sicily, to be the breeding ground of all European eels, thereby claiming the lip-smacking fish and its extraordinary life cycle as the property of the newly united Kingdom of Italy.

But as is the habit of the eel, such hastily grabbed glory quickly slipped out of the Italian's grasp. Grassi had ignored the fact that the thin-heads he'd captured were all almost three inches long. So unless they were born from an unfeasibly large egg, these larval marvels were already quite mature by the time they reached the strait. Could they really have been born so close to Italy's coast? One man didn't think so.

LIKE MANY BEFORE HIM, the oceanographer Johannes Schmidt displayed an almost monomaniacal determination to pin down the obscure breeding ground of the freshwater eel. For nearly two decades the "pathologically ambitious" Dane combed the vast expanses of the Atlantic Ocean in search of freshly hatched fry the size of pine needles. His expedition was so immense and so technically demanding—and led to such an unexpected end—that it would snatch the glory of the eel's slippery story from Italy's grasp and deliver it to Denmark.

His mission began in 1903, when the young Schmidt got a job as a fisheries biologist aboard the *Thor*, a Danish research vessel, studying the breeding habits of food fishes such as cod and herring. One day in the summer of that year, as they were sailing west of the Faroe Islands in the North Atlantic, a puny fish larva showed up in one of the boat's massive, fine-mesh trawls. Schmidt identified the otherwise

insignificant fry as that of the European eel—the first of its kind to be
found outside of the Mediterranean. It was "a chance of luck" that
suggested to Schmidt that the birthplace of the eel was not off the
coast of Italy but approximately twenty-five hundred miles to the
north, unless this thin-head had got very lost indeed.

The Dane became obsessed with locating the eel's true origins,
surpassing even the other eel fanatics—Aristotle, Cairncross, Freud,
Mondini or Grassi—who had been seized before him. Fortunately for
the tenacious scientist, the year before he had made an auspicious be-
trothal to the heiress to the Carlsberg brewery, probably the best lager
company in the world for an aspiring thin-head hunter to hitch his
wagon to, since the company was known to donate generously to marine research.

Suffused with youthful enthusiasm, Schmidt embarked on an al-
mighty quest to locate the smallest possible thin-heads, which would,
he thought, logically lead him to the place of their birth. "I had little
idea at the time of the extraordinary difficulty which the task was to
present," he later wrote. "The task was found to grow in extent year
by year, to a degree we never dreamed of." He trawled fine nets "from
America to Egypt, from Iceland to the Canary Islands," wearing out
four large ships, one of which ran aground near the Virgin Islands and
sank, nearly taking Schmidt's precious thin-head specimens with it.
Then came the First World War. Many of the fishing boats he co-opted
to assist in his mission were shot by German submarines.

Schmidt was also forced to lay siege to an academic establishment
that was infuriatingly reluctant to recognize his painstaking efforts.
In 1912, he had published his first findings: that the farther from the
European coast he traveled, the smaller the eel larvae became, which
indicated that the birthplace of the eel must indeed be in the Atlantic. The Royal Society demurred, however, stating that Grassi's work
identifying the Strait of Messina as the eel's nursery "was considered
sufficient" on the subject, thereby forcing Schmidt back onto a boat
and back out to sea.

A breakthrough came after the war's end on April 12, 1921, in the
southern Sargasso Sea, when Schmidt captured his most Lilliputian

larvae: thin-heads less than a quarter of an inch long, which he presumed were no more than a day or two old. After almost two decades
of searching, the Dane's quest was finally reaching its end. He could at
last feel confident in claiming, "Here lie the breeding grounds of the
eel."

It was an astonishing result; even Schmidt was blown away by the
import of his discovery. "No other instance is known among fishes of
a species requiring a quarter of the circumference of the globe to complete its life history," he wrote in 1923, "and larval migrations of such
extent and duration as those of the eel are altogether unique in the
animal kingdom." Grassi and the Italians had been vanquished, and
the mantle of eel demystification would forever be worn by a satisfied
Schmidt and his Danish homeland.

BUT ONE SHOULD NEVER say "forever" in science or in life.

Nearly a century later, our understanding of the eel's life cycle is
still little more than expensive guesswork. Despite billions of dollars
and the best of modern technology, no adult *Anguilla anguilla* has ever
been tracked all the way on its journey from the rivers of Europe to
the Sargasso Sea. Nor has one been observed mating in the wild. And
no eggs have ever been found.

I asked Kim Aarestrup, a senior researcher at the Technical University of Denmark and one of the world's leading eel scientists, whether
we could be absolutely certain that the European freshwater eel really
is born in the Sargasso. His answer: a sheepish "no."

It's not for want of trying. Modern expeditions have been mounted
attempting to track adult eels with sonar. Researchers have followed
their shadowy deep-sea specters across the Atlantic but have no way
of knowing if they are in fact chasing the right fish—just something
that has the look of it. Still more researchers have fitted hundreds
of eels with state-of-the-art satellite tags. Alas, many of these expensive tags have wound up inside the stomachs of sharks and whales,
their signal relentlessly delivering data to increasingly baffled scientists as the predators journeyed around the ocean, far from the
eels' usual habitats. One wily researcher attempted to trap an eel in

flagrante delicto by suspending traps deep in the Sargasso with seductive females, ripened with artificial hormones so they were practically bursting with the desire to mate. Not even these swollen temptresses could lure adult males into revealing themselves, however. The slippery sirens' cages sank without trace, as did the hope of catching any horny male eels.

Part of the problem is the strange nature of the Sargasso itself. It's breathtakingly deep, over four miles in some of the spots where submarine chasms have formed in the continental shelf. It's thought that the European eel, an ancient species over forty million years old, started breeding in this deep-sea trench when the continents of Europe and America were much closer (geographically speaking). As the continents drifted apart, the eel was forced to migrate farther and farther to return to the place of its birth. The chances of catching it in the act are hampered by these great depths and by dangerous swells—the Sargasso is the only sea that isn't bound by coastline but is instead an over three-million-square-mile whorl surrounded by powerful clockwise currents known as the North Atlantic Gyre. Not only does eel spawning season coincide with the annual cyclone season, but, as Aarestrup pointed out to me, the Sargasso "is slap bang in the middle of the Bermuda Triangle."

That this infamous disaster zone, where countless ships have been swallowed up, is part of the eel's extraordinary story is enough to promote a superstitious belief that Poseidon himself is conspiring to keep the eel's sex life secret.

THE PRIZE FOR CRACKING the eel enigma is no longer just glory, but considerable wealth. Eels are big business. The fish that sustained Mesolithic man may have slipped off the menu in most countries, but they can't get enough of it in Japan, where the humble eel represents a $1 billion annual market. Its fatty flesh is a traditional dish, particularly during the hot summer months, as it's widely believed to have a cooling effect and can help fight off fatigue. Although the Japanese have been known to eat eel ice cream washed down with eel-flavored cola, they mostly like their freshwater eel barbecued and served with a sweet sauce and rice. More than a hundred thousand tons of this Japanese *unagi* are eaten every year. That's a lot of eels to catch.

These days global populations of eel are, however, plummeting, in some cases by as much as 99 percent, due to a combination of overfishing, pollution and other environmental inconveniences—such as massive hydroelectric dams blocking their favorite rivers. The global eel crisis has placed many formerly common freshwater species, including the European eel, on the International Union for Conservation of Nature (IUCN) Red List as a critically endangered species, which makes eating them about as PC as putting panda in your sushi. Although it is hard to drum up sympathy for a slimy, snakelike fish equal to that afforded to a cute-faced, fluffy bear, efforts to get the eel to breed in captivity have been just as intense—if less relished by the press. After billions of dollars and decades of research, the Japanese have enjoyed some success with breeding their freshwater specimen, *Anguilla japonica*, which spawns in a deep trench in the middle of the Pacific. They have found a way to force adults to breed using hormones and even managed to keep a few of the finicky thin-heads alive by feeding them a special diet—of powdered shark egg. Feeding one endangered species with another endangered species' eggs in labor-intensive conditions is not exactly a practical solution. According to Aarestrup, the cost of producing a single glass eel in a Japanese lab is about $1,000 a fish. Which is some pricey sushi.

For now the Japanese must get their eel fix exclusively from dwindling stocks of wild glass eels, caught as they head upriver at the start of their freshwater life and then artificially fattened up in Asian fish farms. Some of these eels hail from Japan or Europe, but most are from America, where few people have shown much interest in eels until quite recently.

The American freshwater eel, *Anguilla rostrata*, is a close cousin of the European; it is also bred and born in the Sargasso Sea, but the larvae migrate to the freshwater rivers of the eastern US seaboard. It's believed to be one of the foods that sustained the starving *Mayflower* pilgrims, after the native peoples generously shared their knowledge of how to catch them. The life-giving legacy of the greasy fish has, however, been gobbled up by the turkey (no celebratory stuffed eels served up at Thanksgiving). A few hundred years later, the freshly inaugurated George W. Bush Jr. pioneered a fashion for wearing specially made eelskin cowboy boots adorned with the blue presidential seal. He took to handing out pairs to his friends—minus the seal but bearing his initials instead (just in case anyone forgot who they came from). But even this lofty endorsement somehow failed to drum up much of a market for the fish in America.

That's all changed now that a glass eel fisherman can earn $100,000 a night by dipping a $25 trap in the right river. The $40 million domestic glass-eel industry has spawned a veritable gold rush in Maine, one of the only places in the United States where elver fishing is permitted. With it has come some slippery business, including shady eel dealers handing millions in cash over to middlemen in motel car parks and fishermen brandishing AK-47s during armed standoffs over the best fishing spots. According to the local press reports, Central American gangs are moving in, as much of the windfall has been spent by fishermen on illicit drugs—although one fisherwoman with an especially lucrative haul was quoted as using her eel dollars to splash out on a new pair of breasts.

IN AN 1879 REPORT about the American eel to the US commission of fish and fisheries, the German marine biologist Leopold Jacoby admitted:

> It is certainly somewhat humiliating to men of science, that a fish which is commoner in many parts of the world than any other fish . . . which is daily seen at the market and on the table, has been able in spite of the powerful aid of modern science, to shroud the manner of its propagation, its birth, and its death in darkness, which even to the present day has not been dispelled. There has been an eel question ever since the existence of natural science.

Not much has changed in the intervening century and more. But time is now running out to crack the eel enigma.

Some experts worry that the survival of the eel is a numbers game. Winning depends on a sizeable population making it back to the Sargasso every year to mate along its idiosyncratic oceanic fronts. If not enough eels show up, they may fail to find one another and simply be swallowed up by this vast whorl of water. If that happens, the eel will have succeeded in keeping its sexy secret all the way to its unfathomable grave.

CHAPTER 2

EAVER

Genus *Castor*

> *There is a very gentle animal called the beaver, whose testicles are extremely useful for medicine. Physiologus says that when the beaver knows the hunter is following him, he cuts off his testicles with his teeth, and throws them before the hunter, and thus escapes.*
>
> —*A Medieval Book of Beasts*, twelfth century

My research has taken me on some strange adventures, but my passion to find the truth about the beaver has probably raised the most eyebrows. It began early one autumnal morning at a rendezvous in a lay-by with a spindly, six-foot man carrying a loaded rifle, complete with silencer, in the boot of his car. His name was Mikael Kingstad and he was a professional beaver hunter.

Kingstad was employed by the city of Stockholm, quite possibly the cleanest and greenest capital I have ever visited. Not far beyond the city's historic, pastel-colored central district, the forests teem with animals, which occasionally spill out of the woodland to explore urban living. Kingstad's job is to make sure these interlopers are kept in check. He's dispatched rabbits (a "problem"), rats (his greatest nemesis), geese ("they produce a lot of shit") and apparently drunken elk. On occasion, he has put overly busy beavers in his sights.

I'm not in the habit of keeping the company of professional animal assassins, but I needed to ask a bona fide beaver hunter an important question: "Have you ever known a beaver to chew off its own testicles and throw them at you?"

OF ALL THE MYTHS about animals, the beaver's perhaps wins the prize for being the most nonsensical. In ancient times this industrious rodent was famous not for its tenacious lumberjack skills, nor for its exceptional flair for architecture, but for its testicles, which were prized by ancient physicians for their medicinal qualities. The savvy beaver, apparently aware of the value of its testicular treasures, was famed far and wide for astute acts of preemptive autocastration in order to avoid death at the hands of the hunter.

The widespread popularity of this unlikely tale is largely thanks to the medieval bestiaries. These early compendiums of the animal kingdom were full of gilded illustrations and earnest descriptions of exotic beasts, from sparrow-camels (ostriches) to camel-leopards (giraffes) and sea-bishops (half fish, half clergyman and whole fantasy). The bestiaries were not the result of some deep commitment to researching the lives of animals, however. Instead they all embellished upon one single manuscript written in the fourth century AD, following the fall of the Roman Empire when the nascent science of natural history had been hijacked by Christianity. The *Physiologus*, Latin for "naturalist," blended folklore with a sprinkling of fact and a heavy dose of religious allegory—a winning formula that turned it into the medieval equivalent of a massive bestseller (only exceeded at the time by the Bible). This archaic manuscript ended up being translated into dozens of languages, spreading absurd animal legends from Ethiopia to Iceland.

The resulting bestiaries are a fabulously bawdy read with much talk of sex and sin, which must have delighted the monks who transcribed and illustrated them for the Church's libraries. They spoke of extraordinary creatures: the weasel, which conceives through its mouth but gives birth through its ear; the bison (or "bonnacon," as it was then known), which avoids a hunter by emitting "a fart so foul that its attackers are forced to retire in confusion" (we've all been there); and the stag whose penis has a habit of dropping off following bouts of

It was a common belief in the Middle Ages that every land animal had a counterpart in the sea: horses and sea horses, lions and sea lions, bishops and . . . sea bishops. This fishy minister, described in Konrad Gessner's *Historia Animalium* (1558), was allegedly sighted off the coast of Poland but looks as if he would be right at home propping up the bar of the Cantina on Tatooine (as featured in the original *Star Wars* movie).

carnal overindulgence. There were more than a few lessons to be gathered and conveyed to flocks of parishioners in such tales. After all, God had created all the animals, and only one—mankind—had lost its innocence. The function of the animal kingdom, in the eyes of the scribes, was to serve as an example for humans. So instead of questioning whether there was any truth to the descriptions in the *Physiologus*, they looked for the human characteristics of animals and the moral values God had hidden in their behavior.

This renders some animals in the bestiaries almost unrecognizable. Elephants, for example, were praised for being the most virtuous and wise of beasts, so "gentle and meek" they were even credited with having their own religion. They were said to have a "great hatred" of mice but a love of country so deep that just thinking about their homeland could reduce them to tears. When it came to fornication, they were "the most chaste," staying with their mates for life—and it was a very long life, lasting three hundred years. They were so averse to adultery that they would punish those they caught in the act. All of which would come as something of a surprise to your average elephant, which enjoys a decidedly polygynous sex life.

The beaver of the bestiaries was a distinctly canny creature whose sagacity doesn't end at tactical self-gelding. The twelfth-century clergyman and chronicler Gerald of Wales wrote: "If by chance the dogs should chase an animal which had been previously castrated, he has the sagacity to run to an elevated spot, and there lifting up his leg, shews the hunter that the object of his pursuit is gone."

This ribald tale of rodent wisdom had a pertinent lesson: man must cut off his vices and surrender them to the devil if he wants to live a life of peace. Such a no-nonsense message of austerity must have delighted Christian moralists, and thus this implausible legend lingered as fact for many centuries.

In the sixteenth century, the great Leonardo da Vinci duly recorded the beaver's astonishing awareness of his gonadal worth in his notebooks: "We read of the beaver that when it is pursued, knowing that it is for the virtue in its medicinal testicles and not being able to escape, it stops; and to be at peace with its pursuers, it bites off its testicles with its sharp teeth, and leaves them to its enemies."

In 1670, the Scottish cartographer John Ogilby was *still* writing about how beavers "bite off their Pizzles and throw them to the hunter" in his rather misleadingly titled tome *America: Being the Latest and Most Accurate Description of the New World*.

What was needed was a dispassionate mind to tackle the truth about the beaver's tackle. Enter the seventeenth-century myth-buster Sir Thomas Browne—who, despite an unhealthy obsession with feeding iron pasties to ostriches, was a lone voice of logic in decidedly muddled times. The Oxford-educated physician and philosopher was the author of *Pseudodoxia Epidemica* (1646), his intellectual assault on what he referred to as "vulgar errors"—the vast suite of popular misconceptions propagated by the likes of the bestiaries that were busy clogging up the emerging discipline of natural science.

This woodcut from the German edition of *Aesop's Fables* (1685) is a fine example of beavers self-gelding in order to pay their testicle tax to the hunter.

Browne's crusade adhered to an unflinching use of "the three Determinators of Truth," which he defined as "Authority, Sense, and Reason," placing him at the vanguard of the scientific revolution by pioneering the modern scientific process. "To purchase a clear and warrantable body of truth," he wrote, "we must forget and part with much we know." And so he began an unsentimental investigation into an array of entrenched untruths ranging from badgers allegedly having legs shorter on one side of their body than the other—an idea he considered to be "repugnant to the course of nature"—to whether dead kingfishers make good weathervanes. (They don't, as Browne discovered after experimenting with a pair of the bird's corpses. He hung them in the air by silk threads and observed that they swung uselessly in opposite directions.)

Browne's nose for nonsense detected something equally fishy about the beaver's testicles. He proposed the fallacy had started in a misreading of Egyptian hieroglyphics, which, for no obvious reason, depicted a beaver gnawing off its gonads to represent the punishment for *human* adultery. This depiction got picked up by Aesop, who publicized the beaver's tale in his popular fables. These in turn were absorbed into the early Greek and Roman scientific literature—and presented as fact. The Roman encyclopedist Claudius Aelianius for instance even credited beavers with employing a neat trick, much loved by drag queens: "Often," he wrote in his epic animal encylopedia, "beavers will tuck away their privates," allowing the savvy rodents to escape while "keeping their own treasures."

Browne assumed the endurance of the tale was largely down to the rodent's quirky biology. Unlike in most mammals, beaver testicles don't bob around outside the body but are instead hidden internally. While these covert "cods" served to validate the idea that the animal had been castrated in some way, Browne rightly pointed out that this anatomical design also meant a beaver couldn't chew off his own testicles even if he wanted to. The ability to "eunachate" was, he said, "not only a fruitless attempt, but an impossible act" that might even be "hazardous . . . if at all attempted by others."

The legend's origins were also, in part, etymological—an area in which Browne had a surprisingly keen eye and ear, since he himself was something of a wordsmith. His clear-headed, protoscientific logic was woven with a florid stream of exotic words, many of which, like "eunachate," he invented himself. Browne is credited with adding almost eight hundred new words to the English language; his coinages of "hallucination," "electricity," "carnivorous" and "misconception" are all still very much in use. Others, like "retromingent"—for urinating backwards—failed to catch on in quite the same way.

Browne astutely noted that the beaver's Latin name, *Castor*, is often muddled up with the word "castrate." The archbishop of Seville was one of many such scribes to have gotten his *Castor*s in a twist. "The beaver (*Castor*) is so named from being castrated," he stated in *Etymologiæ*, his seventh-century encyclopedia.

The Latin *Castor* is not derived from castration but instead connected to the Sanskrit word *kasturi*, meaning "musk" —which brings us to the very nub of the centuries-old fluster over the beaver's cluster.

Beavers were hunted for an oily brown liquid called castoreum that, it turned out, wasn't produced by their testicles, as the legend suggested, but in a pair of doppelgänger organs lurking suspiciously close by.

This case of mistaken testicular identity was first busted a few years before Browne by a French physician and general *bon viveur* called Guillaume Rondelet. Sometime before his death in 1566 from an overdose of figs, this master of dissection took his knife to a couple of beavers and discovered that both the male and the female produced precious castoreum, which they stored in a pair of pear-shaped sacs situated near the animal's anus and connected to its urinary tract. Most mammals have a pair of anal scent glands, which produce a musky substance used to attract mates and mark territory. Rondelet was the first to discover that the beaver is uniquely endowed with a second pair, roughly the size of goose eggs and doing an excellent impersonation of beaver testes.

The similarity of these so-called castor sacs to gonads was so great that other, less keen anatomists had often mistaken female beavers for hermaphrodites and reported mutant males that bore a freaky set of four jewels. But as Browne reminded readers with his characteristic wit, appearances can be deceiving. "Testicles are defined by their office, and not determined by place or situation; they having one office in all, but different seats in many." He diagnosed "the resemblance and situation of these tumors" to be the "the ground of this mistake," and thereby, as far as he was concerned, debunked the beaver myth.

CASTOREUM WAS A REVERED MEDICINE in the ancient world on account of its unusually pungent nature. This was a time when smells were considered to be particularly potent as treatments—the more overpowering, the better chances of being healed. Feces, for this reason, were a firm favorite with doctors, if not their patients. A trip to the doctor might involve inhaling a heady cocktail of up to thirty different medicinal turds (such as mouse and even human), which could only have made the sick feel even sicker. Sniffing beaver "testicles" would have been a bed of roses in comparison.

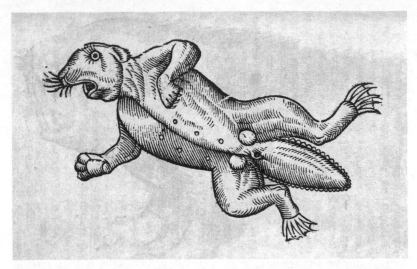

The beaver from Edward Topsell's *History of Four-Footed Beasts* (1607) looks rather taken aback. Perhaps because it is a female, shaved to expose not only her teats but also her "stones," which Topsell tells were much sought after as a cure for everything from toothache to flatulence.

The seventeenth-century British clergyman and naturalist Edward Topsell dedicated several pages of his celebrated bestiary, *The History of Four-Footed Beasts and Serpents*, to castoreum's pungent powers. "These stones," he wrote, "are of a strong and stinking savour." The secretions cured everything from toothache (simply pour warmed castoreum in the relevant ear) to flatulence (don't ask). Their chief use, however, was in the treatment of women's gynecological afflictions. Ancient and medieval pharmacopoeia positively bulged with phallic ingredients; gourds, horns and squirting cucumbers were regularly prescribed for sexual complaints (the tenet presumably being that all every sick woman really needs is a penis, so prescribing her a suggestively shaped vegetable should do just the trick). Beaver "stones" nestled in quite naturally with this phallocentric approach to women's health.

Castoreum was said to completely overwhelm a woman's reproductive organs. The Romans burnt the oily brown secretions in lamps to induce abortions, and, Topsell noted, "a perfume made with Castoreum, Asses dung, and Swines grease, openeth a closed womb." Such was the purported abortive power of the beaver's imposturous balls that it was widely believed that simply stepping over the animal, whether alive or dead, would cause a pregnant woman to lose her baby.

By far the most popular prescription for castoreum was as a tonic for hysteria, a fictitious feminine ailment with a list of symptoms so long they could cater to any hypochondriac: emotional outbursts, anxiety and general irritability being just a few. Hysteria—from the Greek word for the uterus—supposedly arose when a toxic womb wandered about the body, wreaking havoc on a woman's other organs. The nebulous nature of the "sickness" made it a common, catch-all diagnosis for out-of-sorts ladies from Egyptian times onwards. In the seventeenth century, English physician Thomas Sydenham estimated that hysteria was the most common disease after fever, accounting for a sixth of all human maladies. Among women, he wrote, "there is rarely one who is wholly free from it."

Over the centuries, numerous cures for hysteria were suggested. Pelvic massage sounds almost pleasant; exorcism somewhat less so. Inhaling deeply on beaver "balls" was a staple well into the nineteenth century. In 1847, the American physician John Eberle was still pushing the aquatic rodents' anal secretions as the best antidote for hysterical ladies in his *Treatise on the Materia and Medica and Therapeutics*, in particular those with "delicate and irritable habits."

Waist deep in writing this book and feeling a tad hysterical myself, I decided to try and get my hands on the beaver's phony gonads and take a sniff for myself. I wrote a series of surreal emails to hunters I found online, politely introducing myself and then asking them to send me their beaver gland booty. None of them replied.

Next I enlisted the help of a friend who knew her way around the "dark web" and we spent a rainy Saturday afternoon cruising for castoreum. We failed to locate any. In the end, I found a pair hanging out on eBay—a snip (quite literally) at $54.99. Beaver castor sacs, it transpired, are still in demand, albeit not as a cure for hysteria, but as something even more bizarre.

For more than eighty years, the beaver's oily brown anal secretions have been used to add vanilla flavor to a variety of desserts, from fairy cakes to ice cream (ironically, the sorts of food I typically use to self-medicate against feelings of hysteria). How this fact was first discovered boggles the mind, but castoreum is today listed by the US Food and Drug Administration as a GRAS—"generally recognized as safe"—food additive. Fortunately for the beavers, it's not used that often. When it is, the manufacturers need only list it as "natural vanilla flavor" since it is a "natural" product, "naturally" produced by the beaver's nether regions. Which may be enough to squash even the most ardent ice cream habit. Although it's worth noting that one of the most popular synthetic vanilla substitutes contains antifreeze, which makes cooking with beaver anal glands seem positively desirable.

Castoreum also pops up as an essential ingredient in a number of classic perfumes, including Givenchy III and Shalimar. This is less shocking, as the perfume industry has had a longstanding love affair with exotic animal essences. Whale vomit (ambergris) and the genital gland secretions of civet cat and musk deer may sound unappealing, but apparently they are the quintessence of allure.

"The idea of wearing perfumes is to send out a big open-for-business message," scent expert Katie Puckrik explained. "That's what smelling subtly like a critter's hindquarters does for you." By this reasoning, the scent reminds us of our oh-so-sexy, unwashed past. Animal secretions are also valuable for more technical purposes, acting as fixatives for the more volatile ingredients. "They add a sexy 'grrr' to the blend," Puckrik told me, using the trade's jargon. "'Skank' (as we fumeheads call it) provides the bridge between flowers and human skin. Without animalic notes in a perfume, you might as well be wearing air freshener."

My beaver skank arrived a week later. As soon as I opened the Jiffy bag, it was immediately obvious how the legend of the self-castrating beaver had come to pass: out tumbled what looked like a pair of wizened brown testicles. Their smell was seriously intoxicating. Everything that came into contact with them was infused with a strange, woody, leathery odor—not unlike the intense smell of incense that overpowers your average crystal-waving New Age emporium. Edward Topsell claimed that you could spot fake castoreum by the strength of its smell; the genuine article when inhaled should "draw blood out of his nose." The odor emanating from mine was overpowering, but, thankfully, not nose-bleedingly so. It was not at all unpleasant. But then, castor sacs' unusually fragrant nature can be traced to their even more unusual botanical contents.

THERE IS AN evolutionary arms race going on in the natural world, pitting plants against animals that like to eat them. To defend themselves, plants have grown into masters of chemical warfare with the ability to produce a wide spectrum of compounds that range from simply bitter-tasting to the downright deadly. Herbivores breach these defenses by evolving ways to break down, detoxify or recycle these potentially harmful chemicals. In turn, plants up the ante by synthesizing yet more poisons.

Beavers come from a long line of aquatic rodents that have been chewing down trees for construction and feasting on their bark, roots and shoots for at least twenty-three million years. In that time, they've evolved an arsenal of tricks to deal with arboreal chemical weapons, the most inventive being to sequester the toxins and recycle them as their very own defense system. Castoreum contains a multitude of plant compounds: alkaloids, phenolics, terpenes, alcohols and acids, all of which the beaver appropriates from the plants it eats and uses to concoct its very own signature scent. Beavers can recognize neighbors and family from the smell of this chemical fingerprint. They use their castoreum to warn off strangers by marking their territory with the very same chemical message they stole from the plants: bugger off!

Many of these chemicals are also useful to humans. Castoreum's vanilla essence is down to the presence of catechol, an alcohol derived from common cottonwood that is used as a pesticide as well as a food flavoring—a rather unnerving combination of uses, but then many of the forty-five compounds decoded in castoreum have surprising qualities. Phenol from Scots pine has anesthetic properties, benzoic acid from black cherry is used to treat fungal skin diseases and, most crucially, salicylic acid, from the beaver's favorite willow, is the active ingredient in aspirin.

So were ancient doctors correct in prescribing beaver "pizzles" as medicine? Probably not. As far as we know, popping an aspirin is no help at all with combating "the evils," or indeed any of the other illnesses, real and imaginary, that castoreum was alleged to cure. Even if castoreum was a wonder drug, the quantities given to patients were too small to really be effective. I asked a beaver anal gland expert how much I would need to consume to stifle a simple headache and his answer was "A LOT."

In the spirit of Thomas Browne, I decided to give it a go regardless. I waited for a suitably feverish moment and took a shy nibble of one of the sacs that had arrived in the post. It had a distinctly bitter taste that stuck in my mouth for some time, refusing to be dislodged by any amount of toothpaste and scrubbing. About an hour later, I started to belch out potent leathery emissions. The smell seemed to emanate from every pore of my skin and lingered far longer than I was comfortable with. I wound up in the surreal situation of having to meet Dame Shirley Bassey at a BBC recording that evening, overly conscious that I smelled like a beaver's bum and feeling just as fevered as I had before I'd begun my unpleasant experiment.

An eighteenth-century Edinburgh surgeon called William Alexander had a similar experience during a self-administered trial of castoreum's efficacy. He started with a small nibble similar to mine and gradually increased the dosage until he reached eight grams (seriously impressive). Over the week he observed no physiological effect other than "a few disagreeable eructations" (just a few?). He concluded that the stinky panacea was "ill deserving of a place in the current catalogue of medicines."

Now for a dark twist to the legendary tale of the beaver's cunning. According to my Swedish hunter Mikael Kingstad, beavers are sufficiently territorial that if you smear the castoreum of one beaver on another beaver's muddy scent mound, the resident beaver will be compelled to overmark it with its own scent. This has become a handy trick for beaver hunters since all they have to do to catch their prize is smear, sit tight and wait. So it seems the beaver's "pizzles," rather than rescuing the beaver from their attacker, instead became an intoxicating means of luring them towards their death.

This was bad news for the beaver. Centuries of hysterical women had created a huge demand for castoreum. All across Europe beaver populations were being annihilated in order to harvest the smelly elixir. The last beavers in Britain and Italy were killed in the sixteenth century; elsewhere their numbers were in sharp decline. But just as beavers were disappearing from Europe, a whole new continent was discovered, brimming with beavers and heralding a new and even more far-fetched suite of beliefs about them.

"Let us hear no more of the half-reasoning elephant; he is but a ninny to the beaver of America," wrote Frances Thurtle Jamieson in her *Popular Voyages and Travels Throughout the Continents and Islands of Asia, Africa and America*, published in 1820. That this humble rodent's intellect should be exalted above the level of the elephant, an animal whose brain is at least a hundred times heavier, is evidence that early explorers of America lost their minds over the beaver's mental power. Influenced by the folkloric tales of Native Americans, and overly impressed by the beaver's architectural showmanship, they sent home fantastic accounts of an animal Einstein that used its smarts to organize a sophisticated society replete with police, laws and a system of government that rivaled that of man.

Probably the first person to romanticize the beaver in this way was Nicolas Denys, an aristocratic French explorer who sailed to North America in 1632 and went on to become a prominent landowner and politician. He was also among the earliest Americans to detail the natural history of the continent. Of all the animals "of which the industry has been most vaunted," he wrote, "without excepting even the Ape with all that one can teach him," they are all just "beasts" when compared to the beaver—even though this lowly creature, in his opinion, "passes for a fish."

Exactly what kinds of ape or fish Denys had been acquainted with in the Old World are not known, but he described, with extraordinary specificity, his interpretation of four hundred New World beavers uniting as one to build a dam at the start of the summer. They were a highly skilled bunch that walked tall, used their teeth as saws and their tails as hods for carrying mortar or as trowels for plastering walls. There were beaver "masons," beaver "carpenters," beaver "diggers" and beaver "hod-carriers," each doing his duty "without meddling with anything else." Directing this army of skilled labor were a fleet of eight to ten beaver "commanders" who took their instructions from a single overseeing beaver "architect" whose vision dictated where and how to build the dam.

Such unionized industry sounds admirable, but Denys was also keen to point out that this was no bucolic beaver Arcadia. If any beaver was derelict, then his commander "chastises them, beats them, throws himself on them, and bites them to keep them at their duty." To those who might find this portrait of a veritable beaver gulag somewhat hard to swallow, he added, "I would find it hard to believe myself had I not been an eye witness thereof."

Perhaps Denys was limbering up his lying skills for his future role as a politician, because he was by no means telling the truth. When I visited a beaver dam in Sweden with Mikael Kingstad, he laughed hard at Denys's description. Beavers aren't bees. They don't do mass cooperation; they are simply too territorial. Each dam is the property of one beaver family. They build it in order to raise the water level so they can construct a lodge with a permanent underwater entrance, which allows them to stay safely submerged during foraging missions while minimizing their exposure to predators. If another beaver family were to start helping out, the resident beavers would "go ballistic" (Kingstad's words). Even the most impressive dams, which can measure over half a mile long and twice the width of Hoover Dam, are the work of just one family—at most six beavers at any one time—over the course of generations. And although they have been observed walking on hind legs, and carrying their young or twigs between their forepaws and chin, no one has ever seen a beaver using its tail as a builder's trowel.

Nonetheless, stories of the New World were eagerly gobbled up by Europeans back home, and Denys's imaginative depiction of the industrious beaver did the seventeenth-century equivalent of going viral. Endlessly repeated by French colonial travel writers, it inspired a series of highly persuasive paintings that accompanied some of the iconic early maps. One shows fifty-two beavers snaking up a hill in an orderly line, carrying sticks in their arms and mud on their tails, to fashion a dam at the foot of Niagara Falls. It's a charming scene of upright industry until you look closely and read the accompanying legend expanding on the individual beavers' roles. These include "beaver with a disabled tail from having worked too hard" and a menacing "inspector of the disabled," whose job must have been to sniff out those beavers pulling a fraudulent sickie and force them back to work.

As the stories spread, beaver society became ever more fantastic in its organization. "Beavers in distant solitudes build like architects and rule like citizens" announced Oliver Goldsmith in his popular *History of the Earth, and Animated Nature* (1774). The French Jesuit priest Pierre de Charlevoix described the beaver as "a kind of reasonable animal, which had its laws, its government and its particular language."

Beavers were said to always gather in odd numbers so that one beaver may have the deciding vote on matters of beaver democracy. But the whiff of authoritarian discipline was never far away from this rodent republic. "Justice is everything among beavers," warned another French explorer named Dièreville. Those citizens who shirked their civic duties—"the lazy or idle kind"— were "expell'd by the other Beavers . . . just as Wasps are by Bees" to become "vagabonds." According to the French Romantic François-René, Vicomte de Chateaubriand (he of big, bony steak fame), such exiles were stripped of their fur and forced to live out their days in solitary degradation in a dirt hole in the ground, hence they became known as "terriers" from the French *terre,* meaning earth.

The beaver had once again found its life fictionalized and served up as moral guidance, but this time the morals were tailored for a brand new nation founded off the skin of the animal's own back. Beaver fur had become a valuable commodity, used for the broad-brimmed hats then fashionable back in Europe. Hundreds of thousands of pelts were

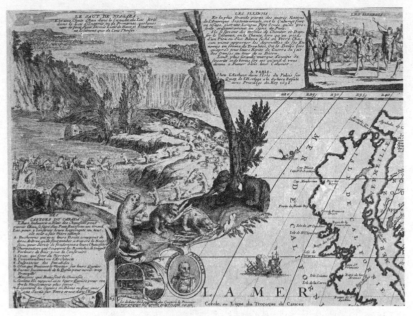

Nicolas de Fer's map of America (1698–1705) provides an imaginative tableau of beaver industry, complete with a handy legend to identify the rodent's roles in this new Utopia. There is, however, nothing to envy about the beaver lying on his back with his legs in the air, identified as "a beaver incapacitated by overwork."

exported every year; the Hudson's Bay Company sold 54,760 at a single sale held in 1763. The beaver's skin even became an official unit of currency in the New World, exchangeable for a single "Made Beaver" coin, which could buy you a pair of shoes, a kettle or eight knives at the local market.

Beaver fever drove the expansion of settlement into the western wilds of America. The beaver was effectively the colonialists' ideal alter ego, its life a parable for righteous living. The rodent's propensity for hard graft, its independence yet willingness to cooperate in order to create public works, appealed to the Puritan ethic. Much greatness could be achieved with a beaver chain gang as your moral guide, but woe betide those who struck out alone to make a quick buck (another unit of currency based on animal skin).

In the Old World, naturalists rushed to interpret the Americans' beaver tales. With the near eradication of its European cousins, beaver architectural wonders and societies were nowhere to be witnessed there. Nor had they apparently ever been. The ancient philosophers had spoken only of the animal's testicles, and thus it was assumed extraordinary industry must be an exclusively New World trait.

The celebrated French naturalist Georges-Louis Leclerc, Comte de Buffon, no stranger to outrageous thinking, took this as evidence of something even more profound. The grandiose Comte was a leading figure of the scientific revolution who strove, somewhat paradoxically, to move natural history out from under the shadow of the Church. His epic, forty-four volume encyclopedia was the first attempt to classify the natural world according to modern scientific principles and remove the superstitious folklore that plagued the medieval bestiaries. It is, however, a hilariously sanctimonious tome, thanks to a delightfully purple prose which, in the style of most science writing of the period, reads more like romantic novel than scientific analysis. Buffon enlisted the beaver as the centerpiece of one of his more extravagant theories, proposing what glory animal societies might achieve if allowed to flourish far away from the corrupting influence of man. "In proportion as man rises above a state of nature, the other animals sink below that standard: Reduced to slavery, or treated as rebels, and dispersed by force, their societies have vanished, their industry has become barren, their arts have disappeared," began Buffon's epic entry on the beaver. In Europe, where beavers had been subject to man's oppressions, "their genius, withered by fear, never again expands." No wonder "they lead a timid and a solitary life." But in the wilds of the New World, where man was a stranger, Buffon believed that beaver society afforded "perhaps, the only subsisting monument of the ancient intelligence of brutes."

The Comte crafted his elaborate theory not just from traveler's tales, but from his firsthand experience of a beaver he'd been sent from Canada in 1758. This beaver lived under his watchful eye at the royal gardens in Paris. After their first year of acquaintance, the French aristocrat was far from impressed by his pet. The beaver was prone to

"melancholy," its "efforts feeble." Although it spent much time gnawing on "the gates of its prison" (and who can blame it), it lacked any
impetus to start construction. The bitterly disappointed Comte was
under the impression that New World beavers were capable of creating elaborate homes, big enough for up to thirty inhabitants and con-
sisting of several floors with windows and even "a balcony to receive
the fresh air, and to bathe." All his beaver had done was mope around
and occasionally bust loose to take a swim in the subterranean vaults
of the gardens. Perhaps the problem was that it was alone, he thought.

The stories of the astonishing power of beaver community, in Buffon's eyes, remained untainted by man and offered hope of a Utopia
we humans could only dream of:

> In this society, however numerous, an universal peace is maintained.
> Their union is cemented by common labours; and it is rendered per-
> petual by mutual convenience, and the abundance of provisions which
> they amass and consume together. Moderate appetites, a simple taste,
> an aversion against blood and carnage, deprive them of the idea of
> rapine and war. They enjoy every possible good, while man only knows
> how to pant after it.

THE TRUTH IS THAT solitary beavers build perfectly well. All Buffon
needed to do was place his beaver near the sound of running water in
order to trigger its industry. The urge to stop the flow is so deeply in-
grained that even a tape recording of a babbling brook will encourage
a beaver to start heaping sticks on top of a loud speaker, despite being
nowhere near any actual water.

This astonishing finding was made by the Swedish zoologist Lars
Wilsson, who spent most of the 1960s basically playing practical jokes
on beavers in the name of science. He raised a number of kits away
from their parents in a man-made environment in order to figure out
whether their dam-building skills were instinctive or learned. By hiding a speaker behind a wall in their enclosure, he discovered the eager
beavers needed only the vaguest audio stimulus to build. He didn't
even have to play the sound of running water; anything similar did

the trick. Even the whir of an electric razor would incite a fury of twig-shoving against the wall by the beavers in a futile attempt to stop the flow.

Such mechanistic behavior makes Buffon's idea of republic-founding beavers look decidedly foolish. But Wilsson's experiment might have made another French scientist, Frédéric Cuvier, younger brother of the famous zoologist Georges, mutter an imperious "I told you so." In 1804, Cuvier became head keeper of the very same Paris menagerie where Buffon had observed his melancholic beaver a few decades before. Cuvier's observations were, however, very different to the Comte's. His new beavers were full of beans and, despite having no parental guidance, busily built away, urged on, in his opinion, by the blind force of instinct.

Cuvier was a follower of yet another famous French scientist, René Descartes, who in the seventeenth century had argued that animals were nothing more than automatons, and only humans were capable of operating with reason. Such unsentimental thinking was now all the rage, a reaction to the overt anthropomorphism of the bestiaries and their ilk. Cuvier held that animal behavior was governed by instinct and that any glimmer of intelligence gradually emerged as you ascended the ranks of mammals: from rodents to ruminants, through pachyderms and carnivores, until one ascended to the lofty pinnacle of the animal kingdom to marvel at the brains of primates, and most especially humans like himself. As a result, Cuvier denied his beavers— mere rodents—any hint of ingenuity.

I spoke to Dr. Dietland Müller-Schwarze, a world expert on beavers, to try to get to the bottom of the animal's brainpower. Müller-Schwarze believes the beaver's extraordinary feats of engineering are mostly achieved instinctively, guided by a set of simple rules, like "build where you hear running water," which we are only just beginning to decode. For example, beavers are often credited with great foresight when it comes to felling trees to ensure they fall towards the water and without becoming entangled in the rest of the forest. "But all the beaver has to do is just make a notch in the tree and let it fall randomly," said Müller-Schwarze. "The chances are that it will fall in the direction

Timber! This beaver's bad day at work is an unfortunate reminder that, no matter how smart you are, not all mistakes can be learned from.

of the open water because the tree grows towards the light, so there will be more branches, and it will be heavier, on that side. It will fall there anyway."

Instinct isn't perfect. The British tabloid press recently featured a photo of a rather unfortunate beaver in Norway flattened to the ground by a tree it had just felled, with the caption "Wood You Believe It?" This demonstrates not only our species' endless capacity for schadenfreude, but also the fact that beavers don't always get it right. On most occasions, however, their mistakes are relatively innocuous and provide a means for learning and adapting their behavior. They have been observed to do both with great resourcefulness, especially when creating and mending their dams. Their capacity to learn, and the complexity of their life skills, may be why beaver kits spend such a relatively long time—more than a year—in the shadow of their parents.

Intelligence is always tricky to assess, but especially so in an animal that is not only nocturnal but spends most of its life underwater or hidden inside an impenetrable lodge. The beaver's physical attributes are far easier to appreciate. Millions of years of evolution have equipped this aquatic architect with the perfect tools for its job: ever-growing, self-sharpening teeth, transparent eyelids that act as swimming goggles, ears and nostrils that shut automatically underwater and lips that can close *behind* their front teeth (allowing for the beaver to gnaw wood underwater without drowning and keeping out irritating splinters when felling trees). Its brain must be similarly well tuned; it's just hard to peek inside the cognitive toolbox.

What we know of the beaver's behavior raises fascinating questions about the boundaries between instinct and learned behavior as well as a rodent's capacity for reason. The beaver may not be prescient enough to chew off its own testicles to save its life or capable of establishing a democratic republic, but even the most conservative ethologist would probably agree with the carefully chosen words of the great animal behaviorist Donald Griffin: "The beaver thinks consciously in simple terms about its situation, and how its behaviour may produce desired changes in its environment."

That's a far cry from Descartes's clockwork creatures.

I'D LIKE TO END with one final beaver tale, a story that perhaps calls into question the intelligence of our own species. By the twentieth century humans had managed to virtually eradicate the beaver throughout most of its range. Across Europe and Asia, populations had fallen to fewer than twelve hundred individuals, who were holding on in just eight small territorial pockets. American beavers were flown in to bolster their numbers and help save the Eurasian beaver.

These introductions were wildly successful, and the new beavers thrived. But then it emerged that American and European beavers are in fact two separate species. They look identical, but the American beaver is more aggressive than its European cousin. The Yankee beaver's bullying behavior was driving the Eurasian beaver even closer to extinction, so it was swiftly declared an invasive species that had to be expunged. But how could government officials, environmentalists and hunters tell them apart? The two species were almost impossible to distinguish without counting their chromosomes.

Then in 1999, Frank Rosell and Lixing Sun from Central Washington University proposed a "quick and easy" way to identify the two beaver species by color-coding castoreum, which varies in tone between the two species. They even developed a handy identification legend for use in the field (bearing an uncanny resemblance to a Benjamin Moore color chart for their most fashionable forest greens and musty yellows house paints).

So in a final twist, the beavers' "testicles" could save them from the hunter's gun, but only if they were first "milked" and their contents carefully color-coded to reveal if they were the right kind of beaver to kill—an ending perhaps even more absurd than the original beaver legend.

CHAPTER 3

LOTH

Suborder Folivora

The degraded species of sloths are perhaps the only creatures
to which nature has been unkind.
—Comte de Buffon, *Histoire Naturelle*, 1749

The sloth is "the stupidest animal that can be found in the world," wrote the Spanish knight Gonzalo Fernández de Oviedo y Valdés, who spent several years exploring the New World and then many more documenting his discoveries in a sprawling, fifty-volume encyclopedia that he published in 1526. The science of natural history was still at the stage where fact was firmly entwined with religion and myth, but Oviedo's main concerns in his tour of the animal kingdom seem to have been steered by his stomach. (The tapir, he said, was "good to eat, and the feet delicious when boiled for twenty-four hours.") Of the sloth's lackadaisical nature, he wrote: "So awkward and slow in movement that it would require a whole day to go fifty paces." He added a menacing note that "neither by threat, blow nor prodding does he move any faster than he is accustomed without tiring."

Oviedo's lethargic lie passed like a game of telephone through numerous travelers' tales. By the time the great buccaneer and wordsmith William Dampier came to write about the sloth in 1676, its pace had slowed to a virtual standstill. "It takes them up to eight or nine minutes to move three inches forward," he wrote with imaginative precision.

"Neither stripes make them mend their pace; which I have tried by whipping them; but they seem insensible, and can neither be frighted or provoked to move faster." Dampier and Oviedo were quite the pair.

I've spent many happy hours watching sloths and can confirm they are mesmerizingly slow, moving as if they are in, or perhaps on, glue. Their average cruising speed is a leisurely 0.19 miles per hour, which is unlikely to challenge even a tortoise, but still not nearly as slow as Oviedo or Dampier suggested. And when frightened they can pick up the pace for short spurts. I've seen a sloth book it up a tree at a surprising lick. But they are physically incapable of going faster than their top speed of .93 miles per hour, since their muscles are engineered to be sluggish—up to fifteen times slower than those of an equivalent-sized mammal like a domestic cat.

To see them in the trees is akin to watching *Swan Lake* in slow motion; they pirouette, sway and dangle with the grace and control of a tai chi master. But if you turn the sloth the "right" way up, gravity removes their dignity. Their limbs sprawl on the ground and they are forced to drag themselves forward by the hooks on their arms, as if mountaineering on a flat surface. This laborious form of locomotion was the main reason why those early naturalists had such a miserable opinion of the sloth—they were observing the animal the wrong way up.

"They are quadrupeds," Oviedo proclaimed, "and on each small foot they have four long claws webbed together like those of a bird." For the record, there is no such thing as a four-toed sloth, let alone one sporting webbed claws, but let's not have such technicalities deter us from hearing out the bold knight. "But neither the claws nor the feet will support the animal. The legs are so small and the body so heavy, that the animal almost drags its belly on the ground."

Sloths are the world's only *inverted* quadrupeds. They have evolved to hook on and hang from trees as the animal incarnate of a hairy hammock. As a result, they have almost dispensed with the need for weight-bearing extensor muscles, like our triceps, which stiffen and protract the limbs. Instead they manage almost exclusively with retractor muscles, like our biceps, that pull them along branches. This

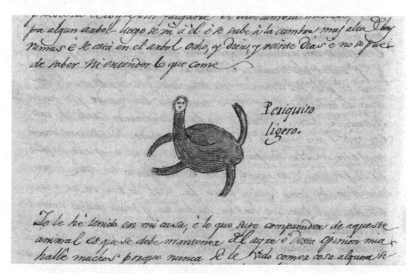

The Spanish conquistador Oviedo was quick to condemn the sloths he found in South America. I'd like to have a word about his drawing skills. I have seen some dodgy illustrations of sloths in my time, but this effort, scrawled among the pages of his sixteenth-century encyclopedia, is, to borrow his own words, "the stupidest thing that can be found in the world."

unusual arrangement demands about half the muscle mass needed to prop up an upright existence, and means sloths can hang out for long periods while expending hardly any energy. It also affords them surprising strength and agility. They can clasp a vertical trunk with only their hind limbs. They can also lean over backwards 90 degrees with freed forelimbs—a trick that, as one sloth researcher pointed out, would, in humans, be "exceptional enough to be shown in a circus."

Oviedo probably never saw sloths in the trees. Instead, local indigenous people probably would have collected specimens and brought them to a village for the Spanish knight's inspection. In this completely unnatural situation, the sloths would have been observed crawling pitifully along the ground. Which is presumably why Oviedo concluded: "I have never seen such an ugly animal or one that is more useless."

The sarcastic nicknames abounded. The Spanish liked to refer to it mockingly as *perico ligero*, meaning "swift so-and-so." Others, like

British cleric Edward Topsell in his seventeenth-century animal guide, referred to "this very deformed beast" as the "Ape Bear."

The Christian church had already begun peddling the concept of "seven deadly sins"—a set of spiritually fatal vices designed to keep the community in check—but sloth, spiritual and physical laziness, had yet to hit the charts. Then finally, in the seventeenth century, after years of arguing, the holy powers that be decided on its definitive seven. Sloth snuggled in at number four, providing Catholic explorers with the inspiration for a catchy new alias for this alien creature. And once they began associating the sloth with sin, all bets were off in terms of a sympathetic understanding of the animal's eccentric biology.

The ensuing tidal wave of abuse reached its grandiloquent peak with George-Louis Leclerc, Comte de Buffon, who offered the first scientific description of the sloth in his encyclopedia, Histoire Naturelle. "As Nature is lively, active, and exalted in the ape species, she is slow, constrained, and cramped in the sloths," he sneered. "Slowness, stupidity, and even habitual pain, result from its uncouth conformation." The sloth, in the Comte's opinion, was the lowest form of animal existence: "One more defect and they could not have existed."

Buffon was writing a hundred years before Darwin's On the Origin of Species rocked the world with its radical idea of natural selection. Nevertheless, Buffon is considered a forerunner of Darwin, and credited by many, including the eminent evolutionary biologist Ernst Mayr, with ushering the concept of evolution into scientific thinking. The abominable sloth, Buffon presumed, had somehow escaped the positive forces that had shaped all other animals towards their singular form of perfection. It was one of nature's "imperfect sketches," a defective remnant that had somehow managed to linger in the animal kingdom despite its inherent "wretchedness." Buffon was the most respected naturalist of his time, and his encyclopedia an international bestseller. The die had been cast. The sloth's destiny as an aberration of evolution was set.

IT IS NO SECRET that I have a soft spot for sloths. I'm fascinated by their freaky biology. But as founder of the Sloth Appreciation Society

I am frequently asked to explain how such an apparently flawed creature could have survived the rigors of natural selection, which ruth-
lessly weeds out the weak. This is when I gather my poise and explain
that sloths are not a defective remnant; they are a group of animals
in reasonable vigor, with six living species in two genera, albeit with
deeply disparaging names: *Choloepus* (meaning "crippled") sloths and
Bradypus ("slow-footed") sloths.

The cripples and the slow-foots are actually as genetically different
as cats and dogs, having diverged from each other on separate branches
of the evolutionary tree some thirty to forty million years ago. Yet
they continue to share the same topsy-turvy, slow-motion life. Such an
adaptation must have its benefits if it was worth evolving *twice*.

The *Choloepus* are more commonly known as the two-toed sloths,
which is something of a misnomer, since they actually have three toes,
but only two fingers. They look like a cross between a Wookiee and a
pig, turned upside down and with hooks for hands. Their long, shaggy
fur ranges from blond to brunette but, whatever their color and despite their cuddly appearance, they are equally bad-tempered. Solitary
in nature, they object to being petted, and if approached by something
unfamiliar, like a hand, they will hiss with mouth open to expose a pair
of quite terrifying-looking teeth. Along with their hooked hands, their
dirty great fangs could inflict a nasty wound if their slow mo swipes
weren't so easy to dodge.

The various *Bradypus*, or three-toed, sloths have both three fingers
and three toes. They sport medieval haircuts and immovable smiles.
They are slightly smaller than their two-fingered cousin—about the
size of a domestic cat under their shaggy, mottled gray and brown
fur—and less cranky but significantly more cryptic. Of the four species,
the largest of the bunch is the marvelously maned *Bradypus torquatus*,
which looks uncannily like a coconut with a mullet. The smallest, and
perhaps most peculiar, are the pint-sized pygmy sloths, *Bradypus pyg-
maeus*; they are less than half the size of other *Bradypus* and only found
in the mangrove swamps of a single island off the coast of Panama.
These dwarf sloths have few natural predators and thus the fabulous
luxury of spending their time grazing on leaves that are believed to

contain alkaloids with a property similar to Valium. So they don't just look stoned, they are stoned.

All sloths are members of the Xenarthra, an ancient superorder of mammals that sound like they're straight out of *Star Trek*, and have the sci-fi looks to match. Other members of this wonderfully eclectic group include armadillos and anteaters. On the surface, this kooky band of misfits appear to have little in common, but closer inspection reveals they all share tiny brains, a distinct lack of teeth and no external testicles. Fortunately, they were named not after these rather unflattering features, but after their other unifying feature—an unusually flexible spine (*Xenarthra* means "strange joints").

This strange superorder owe their oddness to the fact they evolved in isolation from a common relative back when South America was still an island, just after it had torn itself off Africa, some eighty million years ago. Over millennia early sloths flourished in this primeval, forested land, diversifying into more than one hundred species, each filling an unlikely and unique niche. There were giant aquatic sloths that lounged around on beaches and grazed off sea algae; giant burrowing sloths that excavated underground tunnels some 6.5 feet wide; and the most successful of the bunch, the giant ground sloths—the largest of which, *Megatherium*, was the size of an elephant.

Around ten thousand years ago these jumbo sloths all disappeared, leaving behind only the handful of diminutive arboreal cousins that we know today. What wiped out these giant vegetarians has long puzzled paleontologists, who have rifled through caves full of bones and fossilized sloth dung in search of clues. For a while, the consensus was that the last ice age had finished them off. But no; the awkward answer is that we probably ate them.

After South America finally collided with North America some three million years ago and the land bridge was formed between the two continents, some ground sloths ambled north to colonize these inviting new lands. At the same time, rather a lot of humans surged south, carrying spears and smacking their lips at the size of these lumbering lumps of meat. Having survived for many millennia with no natural predators, these defenseless giants ended up on the barbecue.

The other possibility is that they succumbed to our diseases once the land masses converged. Either way we are likely to blame. And it makes sense that the only sloths to survive did so by being small enough to hide in the tops of the trees.

For those, like me, who mourn the loss of these majestic beasts, there is a sliver of hope. Amazon tribal folklore tells of a monstrous creature known as *mapinguari*—taller than a man, with thick matted fur and a fetid odor—that stalks the deepest corners of the jungle. Could this legend betray a lost clique of giant ground sloths roaming the remotest corners of the Amazon? I like to think so.

As a group, sloths have haunted the planet in one shape or another for around sixty-four million years, managing to outlive both the saber-toothed tiger and the woolly mammoth with its stratagem of stealthy living. Of the half dozen species alive today, only the pygmy and maned sloths are considered endangered. That's pretty good going for a lazy loser, and significantly better than other, flashier mammals of comparable size, like the ocelot and the spider monkey. In fact,

The inverted existence of sloths has no need for protractor muscles to hold their limbs erect, so when they are turned the other way up, gravity removes their dignity. Which is why this sloth crossing the road in Costa Rica looks as if it has already been flattened by a car.

one scientific survey from the 1970s found that sloths "are the most numerically abundant large mammal," accounting for almost a quarter of the mammalian biomass—which is a sophisticated biological way of saying you can take your patronizing looks and direct them at some other animal.

"They are survivors," British sloth scientist Becky Cliffe noted. And the very secret to their survival is their slothful nature.

I met Becky while making a documentary in 2011 about the Sloth Sanctuary of Costa Rica where she was working at the time. Sloths may be survivors but they don't get on very well with the roads and power lines that now crisscross their jungle home. Injured adults and orphaned babies are brought to this sanctuary, where they are cared for by self-styled sloth whisperer Judy Avey-Arroyo, who had some rather unorthodox ideas on how to care for wild animals. These included dressing sick babies in bespoke pajamas crafted out of sports socks and keeping a pet rescued sloth called Buttercup (which she referred to as her "daughter") in a wicker hanging chair. It was Becky's job to add some scientific research to all of this worthy sentimentality.

"It's almost as if people don't want to know the truth about sloths. There's something charming about thinking of them as lazy and stupid," Becky said. "As scientists, it's a bit frustrating that's how the sloth has always been portrayed, because we know it's not true."

Becky is determined to debunk the many myths that surround the sloth with sound science and experimental observation. She believes the key to understanding the sloth's slothfulness is its stomach.

The conquistador Oviedo may have thought that sloths "live off air," but the truth is they dine almost exclusively on leaves, many of which have evolved to be tough and often quite toxic in order to avoid being eaten. In the battle between folivore and foliage, the sloth's secret weapon is its massive Buddha-like belly, a multi-chambered monster much like the one found in cows. But sloths are not ruminants and do not chew the cud (regurgitating food while hanging upside down would be challenging, to say the least). Chewing is also not a sloth specialty, since three-toed sloths lack front teeth and the ones in the back of their mouths are little more than ineffective stumps.

Thus sloth stomachs receive barely masticated leaves, which can only be broken down with the help of friendly gut bacteria—a process that needs rather a lot of time.

Exactly how much time was investigated by an American scientist by the name of Gene Montgomery in the 1970s. Montgomery took it upon himself to feed a sloth something it could never digest, namely glass beads, and timed the length of their passage through the alimentary canal until they once again saw daylight. Montgomery waited. And waited. And waited. Just when he thought he had truly lost his marbles, the glass poops arrived, a full fifty days after their journey began.

When Becky repeated Montgomery's study, she chose to use red food dye rather than glass beads. She feared the beads may have skewed the results by becoming lodged in the sloth's system. Nevertheless, her results echoed those of Montgomery's, finding that sloths have among the slowest digestion rate of any mammal.

"For the majority of mammals, digestion rate scales with body size, so larger animals should take longer to digest their food. Sloths appear to break this rule quite spectacularly," Becky said. She thinks it takes the sloth stomach, on average, more than two weeks to break down the cellulose and toxins of a leaf. If it happened any faster, the liver might not cope, and sloths would be in danger of poisoning themselves. The leaves that make up the bulk of their diet deliver little calorific value—the equivalent of a small packet of potato chips a day, about 160 calories. So the sloth has evolved to spend as little energy as possible, performing about 10 percent of the physiological work of a mammal of similar size, which is the very good reason they are nature's couch potatoes.

Sloth bodies have an assortment of nifty modifications to enable their energy-saving, inverted existence. Their blood vessels and throat are uniquely adjusted to swallow food and circulate blood against the force of gravity. Their fur grows in the opposite direction to the norm, with a parting down the middle of their tummy so that rain can run off easily; after a tropical drenching they simply hang out and drip dry. Becky recently discovered that they even have sticky bits on their ribs to keep their stomach, which can be tasked with holding up to a third of a sloth's body weight in slowly digested leaves, from crushing their lungs.

Sloths also maintain a freakishly low metabolism, around half that expected for a mammal their size, and a low core temperature of just 82 to 90 degrees Fahrenheit, whereas most mammals rely on a constantly toasty 96.8 degree internal environment. Rather than keeping themselves warm by stoking their internal combustion engine with calories, sloths wear a thick coat, worthy of an Arctic animal, while inhabiting the warm tropics. Energy from the sun is free, unlike energy from toxic leaves, and sloths take advantage of it by hanging out at the treetops, basking like lizards to soak it up. And like cold-blooded animals, they are able to withstand fluctuations in their body temperature of several degrees throughout the day.

"They are very economical animals and they make the most of every single thing they have available to them," Becky said. "If they have to and they are pushed to the extremes, then they are more than capable of pushing through and using alternative ways. Like their metabolic rate—it looks like they can raise their metabolic rate and heat themselves . . . if they absolutely have to. But most of the time they don't need to."

It has long been asserted that sloths completely lack the ability to control their body temperature, which helped contribute to the idea they were somehow less evolved than other warm-blooded mammals. Becky has refined the research on sloths by incorporating modern techniques and equipment. To study sloth metabolism, for example, she used a custom metabolic chamber and a "bum-o-meter," which she "lubed up" and inserted into an accommodating sloth as part of her experiments. (The extraction of a sloth's secrets may not always be dignified.)

"Every time I get a new piece of data it just shows me more and more that sloths are not as incapable of surviving as we think they are," Cliffe said. "They've been around for sixty-four million years . . . If they were completely incapable of raising their own body temp they'd have died out a long time ago."

G. Edwards ad riv. delin.　　　J. M. Seligmann excudit.　　　Joh. Seb. ast. Leitner sculps.
Cum Priv. Sac. Caes. Majestatis.
Ignavus.　　　　　Nᵒ 100. VIIIͭᵉʳ Theil.　　　Le Paresseux.

The illustrators of the early natural history books churned out some fantastically bad drawings of sloths, which were often quite humanlike in appearance. This one from the 1770s by George Edwards and Mark Catesby has more than a touch of hippie about it (apart from the Freddy Krueger claws).

Cliffe suspects the sloth's low metabolism could be the cause of what she called the ultimate sloth survival superpower: its seeming ability to cheat death, another rumor that has been the subject of much speculation over the centuries, but this one might have a bit of truth to it.

Back in 1828, a British naturalist called Charles Waterton noted, "Of all animals, this poor ill-formed creature is most tenacious to life. It exists long after it has received wounds which would have destroyed any other animal." There have been reports of sloths falling almost a hundred feet to the forest floor without injury, of surviving forty minutes immersed underwater and of lasting twenty-four hours in a fridge. One sloth was even said to have lived for thirty hours following decerebration.

Many of these stories were certainly exaggerated, but over the years, the team at the Sloth Sanctuary in Costa Rica had seen animals make near-miraculous recoveries, whether they'd been zapped by power lines, attacked by dogs or hit by cars.

"Why and how they are capable of bouncing back after horrible injuries is still a bit of a mystery," Cliffe said. A conversation with Professor Enrique Amaya of the University of Manchester, who specializes in gene expression and limb regeneration in gecko lizards, provided her with a fascinating lead. "When a gecko needs to regrow its tail, it enters what he calls an 'embryonic state'—basically, it lowers its metabolic rate in order to put all of its energy into healing." She suspected that the sloth's low metabolic rate might work in a similar fashion, but so far she hasn't been able to back up her hypothesis with experimental evidence.

Others have speculated that the sloth's slow metabolism might protect it from cancer, or aid evolution by allowing birth defects to persist and develop into beneficial new structures—like its unusually long neck, which has more vertebrae than any other mammal's, even the giraffe's. The trait appears to have evolved from extra rib cage bones, which were co-opted into neck vertebrae—the kind of large-scale deformity that's normally wiped out by a mammal's immune system but which survived in the sloth as an adaptation. The long neck allows nature's couch potato to turn its head 270 degrees and graze leaves all around it without wasting precious energy moving the rest of its body.

THE SLOTH'S SLOW METABOLISM doesn't mean the creature is sleeping all the time. Although it was long reported that the so-called world's laziest animal spent around twenty hours a day snoozing, a recent study by Niels Rattenborg and Bryson Voirin of the Max Planck Institute found that sloths in the wild actually sleep less than half that amount, an average of 9.6 hours a day.

"Just because they are not moving doesn't mean they are asleep," Becky told me with the authority of someone who has spent a lot of time living alongside sloths. "*Bradypus* in particular don't sleep like other animals do, for nine or ten hours at a time." Unlike other mammals *Bradypus* apparently lack a daily rhythm; they are neither nocturnal nor diurnal but compulsive nappers instead. The majority of their waking day (and night) is spent quietly hanging in the trees in a seemingly meditative state, motionless, with their eyes open and staring blankly into space. This wakeful but inactive state comprises the majority of their existence. It's crucial for energy conservation, and their ultimate survival.

SO HOW DOES A virtually immobile, somnolent bag of fermenting leaves avoid getting itself eaten?

The sloth's main predator is the harpy eagle, an animal so terrifying it is named after the wind spirits of ancient Greek mythology who carried the dead to Hades. The harpy is one of the world's biggest and fastest raptors, with talons the size of grizzly bear claws and an over six-foot wingspan. It can fly at speeds of up to eighty miles per hour. It boasts razor-sharp eyesight and a ring of feathers around its face that focus sounds so that it can detect the slightest rustle of a leaf.

The sloth doesn't seem like much of a match for this apex predator. With its poorly developed ears and eyes, it lives its life in a muffled blur. These severely blunted senses are unlikely to provide advance warning of a savage, winged assailant. And with its top speed at just under a mile an hour, running from danger is clearly not an option.

Sloths' apparent lack of defense was one of the Comte de Buffon's big beefs with the species. "Sloths have no weapons either offensive or defensive. They have no cutting teeth, the eyes are obscured with hair," that "resembles withered herbs."

Within that much maligned coiffure lies the sloth's secret weapon. The sloth, you see, is a master of illusion, capable of disappearing into the rainforest behind a cloak of invisibility. Its coat is a miniature eco-system that rivals Edward Lear's old man with a beard. Special grooves in the fur collect water and act as hydroponic gardens for eighty different species of algae and fungi, giving the sloth a greenish hue. It also supports a wealth of insects. One study by Jeffrey Waage found a single sloth playing host to nine species of moth, six species of tick, seven species of mite and four species of beetle—including 980 individuals from just one of those beetle species. (For the nit-picking scientists out there, three of those mite species were, strictly speaking, anus-dwellers.)

Crawling with bugs and looking as if they have been dragged through a hedge backwards isn't going to win the sloth any beauty contests, but it means they look and smell exactly like a tree. And most of the time they are as motionless as one too. When they do move, they don't crash about like monkeys; their arboreal ballet is as silent as the gentlest breeze and so slow it is thought to slip under the radar of the monstrous harpy as it swoops above the canopy, scanning for prey.

The first person to appreciate the sloth's stealthy strategy for survival was the American naturalist William Beebe, an advocate of studying animals in their native habitats and widely considered the father of field ecology. Beebe devoted most of his life to a series of increasingly dangerous expeditions. He circumnavigated the globe to document the world's pheasants—a study that briefly cost him his sanity and permanently his marriage. He also probed the depths of the ocean more than thirty-five times in a precarious-looking metal eyeball he called the bathysphere, becoming the first human to dive to depths beyond three miles. He was still shinning up slippery tropical trees to get a closer look at birds' nests well into his eighties.

Here we see the father of field biology, William Beebe, limbering up to hurl one of his sloth study subjects into the river in order to check out its backstroke.

Beebe spent much of the 1920s observing sloths in the wilds of British Guiana, an experience that led him to consider rather than condemn the curiosities of this creature. He ridiculed the Comte de Buffon for his antiquated, narrow-minded views. "A sloth in Paris," wrote Beebe, "would doubtless fulfill the prophecy of the French scientist, but on the other hand, Buffon clinging upside down to a branch of a tree in the jungle would expire even sooner."

Many of Beebe's findings are still quoted today, even if some of his methods were a little, shall we say, *alternative*. He was, for instance, among the first to report that sloths can swim, something he discovered after repeatedly chucking one in a river.

Their unique digestive system produces a surfeit of gas, and evolution has found a suitably resourceful use for this excess of trapped wind. It acts as a built-in, biological buoy. Sloths can actually bob along three times faster in water than they can move on land, using their long arms to do a half decent doggy-paddle. According to M. Goffart, author of the definitive *Function and Form in the Sloth*, if you flip them over, they can even do passable backstroke.

When Beebe wasn't throwing sloths into the water, he was shooting at them. "I have fired a gun close to a slumbering sloth and to one feeding and aroused but little attention." From which he concluded that sloths are not so much deaf but show a "general disinterest in the noise." They didn't even react to the sound of a predator such as a nearby crying hawk; "neither sight nor sound penetrated the dull aura of mental opacity which invests the senses of these mammals."

I know this from my own experience shouting "Boo!" at a sloth. The only response, if any (yes, I did it more than once), was a much delayed and distinctly dreamy turn of the head. The fact that it is impossible to startle these masters of mellow could be yet another cunning camouflage tactic. Jumping out of your skin every time you see a harpy eagle is an impractical reflex response for a creature wishing to stay concealed.

SLOTHS ARE LARGELY SOLITARY characters, so their apparent deafness may also be down to the fact that they have dispensed with virtually all forms of vocal communication except one: female *Bradypus* will scream for sex. A lusty female sloth will climb to the top of a tree and let out an ear-piercing shriek that travels several miles to advertise her window of fertility. We have Beebe to thank for the precise note of this yodel—D-sharp. No other note had an effect on the males. "They are attuned to this sound and only this sound. I gave a C and E, then upper B with no result whatsoever. Again I whistled D-sharp and the reaction was as instantaneous as a sloth can achieve." This shrill whistle, noted Beebe, perfectly mimics that of a kiskadee flycatcher—another sneaky sloth adaptation for remaining hidden, even when a female is broadcasting her location at the top of her lungs.

Beebe may have heard their sexy screeches, but he never saw sloths mating. Despite their high-pitched courtship yodeling, sloths manage to keep their actual sex life on the down-low, making it prone to much mythologizing. There is a persistent rumor on the Internet that sloths are so slow that it takes twenty-four hours for them to have sex. This isn't true. As the first person to film wild sloths mating, I can report that sloth sex is a surprisingly swift and athletic affair. The male approaches the female and, after a short display of flamboyant posturing, the deed is done and dusted in a matter of seconds. It appears sex is the only thing sloths do quickly.

This may not sound wildly romantic, but it makes a great deal of sense. The movement of mating exposes the sloth's position to predators, so it's a good idea to get it over with, *tout de suite*. Prolonging the act would also waste precious energy. That said, the sloths I observed were at it repeatedly over the course of an afternoon, mating every half hour or so, with the male retreating to snack on a cecropia leaf and take a deep snooze between each bout.

THE BIGGEST CONTROVERSY in sloth biology is perhaps even more intimate: it concerns the sloth's curious defecation customs. The ordinarily idle folivore has a rather perplexing habit of descending all the way to the ground in order to relieve itself. It's a lengthy and highly ritualized affair, with the sloth hugging the base of the tree and wiggling its bottom on the ground to neatly dig a hole with its stumpy little tail in which it can do its business. Afterwards, the sloth has a good sniff and then neatly covers its work with leaves, before making the long commute home. Five to eight days later, it goes through the routine again.

This extravagant ritual is such a key part of sloth behavior that the sanctuary in Costa Rica potty-trained orphaned baby sloths at designated "poo poles" stuffed in the ground around the front lawn. The woman in charge of teaching them was a distinctly unslothlike character called Clare Trimer, who'd spent most of her life working with the inmates of a high-security US prison and positively vibrated with nervous energy. She'd found her own sanctuary with the sloths in an early retirement, but she treated her job of teaching the babies the "poo dance" with the intensity you'd expect from someone accustomed to the stressful conditions of a correctional facility. Clare showed me how you could identify when a sloth was actually on the job by the "blissed out" look on his face. That famous sloth smile "gets a little bit wider and they kind of zone out."

Sloths' bathroom rituals may be satisfying, but they come at a cost: the laborious expedition to the forest floor is energetically expensive and can be quite perilous. Forsaking the cover of the canopy strips them of their invisibility cloak, exposing them to the sharp eyes of ground-based predators such as the jaguar. More than half of sloth deaths are estimated to occur while they are doing their business. For these animals that are so perfectly adapted to a life hanging in the trees that they are born there, mate there and even die there, it is strange they don't also simply defecate there, as monkeys do.

This pooping problem is a source of vigorous debate among sloth researchers. Staff at the Sloth Sanctuary told me sloths come to the forest floor to fertilize their favorite tree. That's a lovely, hippy ideal, but sloth dung, having taken around a month to digest, returns to the forest as highly compacted cellulose bricks that make for poor compost. Others have argued sloths come to the ground to eat dirt in order to absorb minerals missing from their leafy diet, a hangover from their days as giant ground sloths. This theory has thus far failed to get much traction.

In 2014, a pair of American ecologists, Jon Pauli and Zach Peery, made a big splash when they claimed to have solved the scatological mystery once and for all. The answer, they said, was hidden in the well-established relationship between the sloth and a moth that lives exclusively in the sloth's fur. Wild sloths are riddled with these drab little insects, which creepily crawl over their faces and rise in fluttering plumes of silvery-gray wings when disturbed. The moths' life cycle appears to be intricately tied to the sloth's bizarre bathroom behavior. The larvae are *coprophagous*, or, put more bluntly, dung-munchers. So the adult moths lay their eggs in the sloth's excrement, and once the larvae metamorphose, they fly up into the trees to hitch a ride on another sloth on its way to the loo. And thus the cycle of life continues endlessly.

The theory the US ecologists suggested is that these moths were little more than "flying genitals," since their brief adult life is all about sex; as soon as they mate they die. As a result sloth fur is chock-full of both living moths and decomposing ones, the latter of which the ecologists believed serve as fertilizer for the algae growing in the fur. This could very well be true, but their theory did not end there.

The Americans pumped the stomachs of two- and three-toed sloths and found they contained algae. From this evidence they concluded that sloths must be grazing off their own fur to supplement their low-calorie, mineral-deficient diet. This led to an imaginative leap that the sloths must therefore be risking their lives in order to perpetuate the life cycle of the moth because it fertilizes their "algae gardens," which would make sloths the most committed farmers in the entire animal kingdom, risking death for the health of their crops. The fact that no one has ever seen a sloth licking or eating its fur did not deter the ecologists' enthusiasm. They simply presumed the sloths were doing it at night, or in secret.

When the ecologists' paper was published, it generated a lot of attention in the media: sloths and eccentric pooping habits are *made* for a slow news day. The scientific community were a touch more skeptical about the sloth's alleged surreptitious midnight feasts. "It's a very unfortunate conclusion," said Becky Cliffe. "Anyone who has observed sloths in the wild for any amount of time knows this is not the case. It paints a picture that they hadn't spent much time in the field observing wild sloths themselves. The fact that sloths have this algae in their stomachs simply says to me that the algae must exist in nature, that they ingest it some other way."

If the sloths were simply providing the moths with a place to reproduce, why would they need to return to the base of the same few trees? Cliff had set up surreptitious poo-cams to catch their habits. "Wild sloths always poop in the same places. In fact, multiple poo piles can be found at the bases of certain trees," she said.

Cliffe has her own theory as to why the sloths make these risky loo runs: she thinks it's all about sex.

The penny dropped for her while she was conducting her red-dye digestion study. As part of the work, she collected the feces of several subjects at the sloth sanctuary, and then—oh the glamour of the scientist's life—stored it in her bedroom, since her makeshift jungle lab was also the hut where she lived. A visit to Cliffe generally required picking your way across a floor strewn with dozens of dung piles, each heaped on a piece of A4 paper with enigmatic scribbles on them, like "Brenda, Day 4."

One night, while sleeping among her dung heaps, Becky was woken by a "tap, tap, tap" at the window. She drew back the curtain and was shocked to see a male sloth staring back at her, apparently trying to break into her bedroom. When she left her hut in the morning, he was still there. The next night he came again, making a slow dash for the inside of her hut when she opened the door. After several days of being stalked by this persistent sloth, Becky realized what he was after: Brenda's poo.

Brenda had been screaming for days at the time of her fecal collection. This led Becky to conclude that sloth dung must contain pheromones and their latrines must act as message boards, as they do for many other mammal species. Depositing a humongous dump of fibrous dung is therefore the sloth equivalent of a personal ad. Females come down to the ground to advertise their location, readiness for sex and a host of other personal details, as well as to check out the competition—and so do males. Which makes the sloth's toilet trips a bit like speed dating, but without the speed.

"It makes sense—a big risk like coming down from the trees requires a big payoff, and the biggest payoff of all is reproduction," Becky said.

If Becky's right, this clandestine communication system is yet another way the sloth blends in with its rainforest home to keep its existence secret—the motivating drive of their lives that has made them so successful in their native habitat and so hard for flamboyant bipedal apes like us to understand.

CHAPTER 4

HYENA

Species *Crocuta crocuta*

The hyena, hermaphroditic self-eating devourer of the dead,
trailer of calving cows, ham-stringer, potential biter-off of
your face at night while you slept, sad yowler, camp-follower,
stinking, fowl, with jaws that crack the bones that the lion
leaves, belly dragging, loping away on the brown plain,
looking back, mongrel dog-smart in the face.
—Ernest Hemingway, *Green Hills of Africa*, 1935

The hyena has been censured by more scandalous untruths than even the sloth. They are considered nature's thugs—condemned throughout history and across cultures and continents as dim-witted cowards, skulking in the back alleys of the animal kingdom, waiting for an opportunity to mug other, more noble, animals of their dinner.

The largest, most widespread and most misconstrued of the four species of hyena is the spotted hyena, *Crocuta crocuta*. With its scrappy fur, hunched back and wide drooling grin, this so-called laughing hyena may not be the prettiest of animals. But our disdain goes more than skin deep; it's personal. The hyena's bewildering biology is partly to blame, which caused no end of consternation for taxonomists. Here is an animal that looks and hunts like a dog but is in fact a souped-up member of the mongoose family and therefore more closely related to a cat. Over the course of several editions of his tome *Systema Naturæ*,

Topsell's encyclopedia was very confused about the hyena. It was perhaps a bear, or a dog, but its short tail reminded the cleric of an ape, and it had the feet of a woman. He also claimed that hyenas continually held up their short tails in order to flash their backside—flaunting their hermaphrodite nature.

Carl Linnaeus classified the hyena first as a cat, then later as dog. He never got it right.

Others labeled the hyena as some kind of hybrid—a serious denunciation since, according to Sir Walter Raleigh's analysis, it made them unfit to travel on Noah's Ark. In his seventeenth-century classic *Historie of the World*, Raleigh had wrestled at great and earnest length with the issue of how to fit the entire animal kingdom, Noah's family *and* enough food for all onto the Lord's lifeboat. It would, he hypothesized, be quite some squeeze. His "rational," space-saving solution was to leave hyenas (the unholy offspring of foxes and wolves) to drown: "For those Beasts which are of mixt natures," he wrote, "it was not needful to preserve them; seeing that they might be generated again by others."

EVEN MORE BAFFLING than what kind of animal the hyena is, was the basic question of its gender. "It is the vulgar notion, that the hyæna possesses in itself both sexes, being a male during one year, and a female the next," wrote Pliny the Elder in his animal encyclopedia.

The Roman naturalist wasn't the first—or the last—to suggest the hyena was a hermaphrodite, capable of swinging its sex according to the seasons. The rumor was rife in African folklore and debated by Aristotle. And hermaphrodites aren't unknown in nature. Many species of worm, slug and snail are both male and female; there's even a whole class of bony fish that can switch between the two sexes willy-nilly. Indeed, modern science has identified more than sixty-five thousand known cross-sexual species. The hyena is not one of them.

The likely stimulus for Pliny's sexual mythologizing is the female spotted hyena's wildly unconventional genitalia, which are a near perfect facsimile of the male's. A *Crocuta*'s clitoris extends almost an impressive eight inches and is shaped and positioned exactly like a penis, hence known in polite biological circles as a "pseudo-penis." The original chicks with dicks can even get erections. To complete this trans trickery, the female spotted hyena also appears to sport her very own pair of testicles: her labia have fused to form a false scrotum and are filled with fatty-tissue swellings, which are quite understandably mistaken for male gonads.

Paul A. Racey and Jennifer D. Skinner, authors of a scientific paper on sexual mimicry in the spotted hyena, declared the appearance of males and females "so close that sex could only be determined with certainty by palpation of the scrotum." Copping a feel of a spotted hyena's soft bits would seem a rather foolhardy pastime for anyone wishing to hang on to their hands, but it certainly helps to explain Pliny's slipup: he was a prolific plagiarist and highly unlikely to have ever clapped eyes on a spotted hyena's genitals, let alone fondled them. It wasn't until the late nineteenth century that the British anatomist Morrison Watson got his hands on a hyena's nether regions and the hermaphrodite rumor was finally crushed. Fortunately for Watson, he survived his intimate encounter.

Today, the female spotted hyena is the only known mammal with no external vaginal opening. Instead, she must urinate, copulate *and* give birth through her strange, multi-tasking pseudo-penis. This last eye-watering feat is like squeezing a cantaloupe out of a hosepipe, and one in ten first-time hyena mothers die in the process. The fate of their cubs is even more precarious, since the umbilical cord is too short to navigate a birth canal that's not only twice the length of a similar-sized mammal's but includes a cheeky hairpin turn halfway down. Up to 60 percent of cubs suffocate on their way out.

It is easy to see how the sight of a "male" hyena giving birth through his penis led to the hermaphrodite myth (as well as a few lingering nightmares), yet with so many fatalities during reproduction, it's less easy to see why the hyena's pudendum took such a peculiar evolutionary path in the first place.

The female's gender-bending doesn't end with her fake phallus. Spotted hyenas are unlike all other mammals in that the females are significantly bigger than the males and much more aggressive.

"You do not want to be a male spotted hyena," Kay Holekamp, a professor of evolutionary biology and behavior at Michigan State University, told me. She's spent over thirty years studying *Crocuta* in the wild, painting a fresh portrait of these miscast creatures and earning a reputation as the Jane Goodall of hyenas.

Every hyena clan is a matriarchy ruled by an alpha female. In the clan's strict power structure, dominance passes down the alpha female's line to her cubs. Adult males rank last in the hierarchy, reduced to submissive outcasts begging for acceptance, food and sex. At a communal carcass, where thirty or so hyenas might be vying for their pound of flesh, adult males eat last—if there's anything left—or risk violent retribution from the sisterhood.

Holekamp believes the driver of the female spotted hyena's aggression and dominance is this intense competition over carcasses. A frenzied scrum of hyenas can turn a 550-pound adult zebra into a bloody stain on the grass in under thirty minutes. An adult can gobble up to a third of its body weight per feeding—between 33 and 44 pounds of meat. It's a frantic, frenzied and, at times, frightening scene. A female that is bigger and meaner has a better shot of ensuring her surviving cubs get a place at the table and don't get hurt in the process.

Dominant females have another trick for giving their cubs an aggressive advantage. A recent study by Holekamp has shown that the more powerful the female, the more testosterone her fetuses are exposed to during the final stage of pregnancy. These androgens are produced by the mother's ovaries, which is unusual enough. It is Holekamp's belief that female cubs are more sensitive to their effects than males are. Spotted hyenas have an unusually long gestation period, and marinating in this prenatal androgen bath affects the development of the cub's nervous system so that they are wired to fight from the moment they are born. And they already have the necessary weapons: unlike most mammals, hyena cubs emerge with eyes open, muscles coordinated and teeth already pierced through their gums and eager to bite. These belligerent newborns frequently fight to the death over dinner, and siblicide is commonplace.

Scientists presumed that this superfluity of prenatal testosterone also caused the disproportionate growth of the female hyena's clitoris. But when the same researchers gave captive pregnant spotted hyenas a diet packed with anti-androgens (which blocked the male sex hormone), female cubs still surprised them by popping out of the birth canal sporting "a large pendulous phallus" and a "normal pseudo-scrotum."

According to Holekamp, the spotted hyena's extraordinary sexual equipment remains "one of the most interesting mysteries in biology." Some scientists have suggested the hyena evolved the pseudo-penis in order that it could be licked by subordinates, which is how female spotted hyenas tend to greet each other (and assess dominance). But as appealing as this theory may be to the scientists championing it, Holekamp can't see how this would be a strong enough evolutionary driving force to create a structure as reproductively damaging as the pseudo-penis. "I am convinced we can rule out all the hypotheses that have been forwarded in the literature to date: it is definitely not merely

a 'side-effect' of female androgen exposure, and it is not there to permit greeting behaviors," Holekamp said.

Holekamp's educated guess—and it is still a guess—is that the female's gender-bending is a result of the age-old war between the sexes.
Unlike most animals, where the males duke it out and the winner gets
the girl, in spotted hyena clans the females dictate the who, where
and when of copulation. Sex is an undignified affair that sees the male
forced to squat at the female's rear, stabbing away blindly in an attempt
to insert his erect actual penis into her floppy, half-foot pseudo-penis.
It's a bit like the male's trying to have sex with a sock—a pretty tricky
exercise that's completely impossible without the female's full cooperation. Brute force alone simply won't work, as it does for the males of
other mammal species like dolphins, among which nonconsensual sex
is surprisingly commonplace. The female hyena's pseudo-penis may
be acting as an "anti-rape" device, allowing her to exercise choice over
whom she mates with.

This is pretty handy because, in addition to the dangers lurking in
her precarious birth canal, a spotted hyena suffers a few other reproductive challenges. Her ovaries have comparatively little follicular tissue and produce relatively few eggs, and so it pays for her to be picky.
You'd never imagine this was the strategy from observing her behavior, since female spotted hyenas are highly promiscuous. Holekamp
reckons the pseudo-penis allows the female to choose not just whom
she mates with but, more impressively, who actually fertilizes her precious eggs by acting as a form of built-in birth control. That strangely
elongated reproductive tract, with its various twists and turns, slows
down sperm as they swim towards their goal. If the hyena changes her
mind about a male after mating, she simply flushes out his semen by
urinating. Go, sister!

I WONDER WHETHER this new image of the spotted hyena as a pioneering feminist, strutting around the savannah with a counterfeit
cock, beating up on submissive males and taking control of her sexual destiny, would have been any less sacrilegious to the male bestiary
authors than the original hermaphrodite myth. In the hands of these

Sex for a male spotted hyena is no laughing matter. Inserting his erect penis into the female's flaccid pseudo-penis is a challenge akin to one of those fairground games that you can never win. No wonder this brave male has attracted a spectator, to see how it's done.

religious scribes, the animal's suspicious sexuality rendered it "a dirty brute," and hyenas were frequently used to warn against the evils of homosexuality.

This set the stage for a resolutely damning portrait of the hyena that often included grisly tales of grave robbing. This ghoulish hyena habit was first put forward by Aristotle, but the bestiary authors were emboldened to embellish such tales to service their moralistic needs. The hyena was said "to live in the sepulchres of the dead and devour their bodies." The animal was a terrible fiend "without pity for the living and ominous to the dead."

The myth of grave robbing persisted all the way into the nineteenth century. The Victorian naturalist Philip Henry Gosse was inspired by the hyena to pen particularly purple prose that owes more to Mary Shelley and the fashion for Victorian Gothic horror than it does to the truth. "In the Place of Tombs, gleam two fiery eyes," he wrote in

racione fuftenram ; ᴅᴇ ᵽ ᴇ ᴎ ᴀ Hiena

uralıquando mafculuf fır alıquando femına.

Most medieval bestiaries featured a graphic illumination of a hyena, back arched and teeth bared as it busted into a tomb to savage a human corpse—a striking image that ensured their negative branding.

1861, in his massively popular *Romance of Natural History*, "with bristling mane and grinning teeth, the obscene monster glares at you, and warns you to secure a timely retreat." Other naturalists of the era showed a tad more restraint, but they still described the hyena as "a most mysterious and awful animal," "rank and coarse" with "revolting habits." This creature, they decided, was "adapted to gorge on the grossest animal substances, dead or alive, fresh or corrupted," and as such was "cordially detested by the natives in all countries."

Kay Holekamp told me that spotted hyenas in East Africa have indeed been known to dig up human corpses. "Their sense of smell is so acute they can even detect food underwater," she said. The Nuer of Sudan have a saying that the only way to heaven is through the gut of the hyena. And some African tribes encourage hyenas to consume their dead by covering corpses in fat and leaving them outside where they can be easily taken—a tradition that Western missionaries worked hard to curtail. But Holekamp insists hyenas only exhibit such behavior when "times are tough." The perpetuation of this gruesome legend is more a mark of our disgust for any animal scavenger.

Western society tends to admire animals that are perceived to do a hard day's work, like beavers, or who make their meals by hunting or gathering. But scavenging is an honorable profession, one that recycles energy and prevents the spread of disease.

And the hyena is *very* good at its job. It is the garbage truck of the African plains, using its powerful jaws and stomach acids to digest what most animals can't. The fact that it fails to get sick after gorging on a putrid carcass riddled with anthrax may explain why many cultures believe the hyena possesses magical powers.

In terms of tonnage of meat consumed, hyenas are the most significant terrestrial carnivore on the planet. However, only the brown hyena and the striped hyena are primarily scavengers. Spotted hyenas are highly efficient predators, killing 95 percent of the food that they consume. Hunting parties are capable of bringing down aggressive animals, like water buffalo, several times their size. Even lone hyenas have been known to catch impressively large prey—one bold strategy being to lock on to the animal's testicles and hang on while dodging a battery of defensive hooves until the victim bleeds out.

Such tactics are not for the faint-hearted, yet somehow hyenas have an enduring reputation for being a bunch of wimps. "All writers agree that the hyaena lacks courage," wrote the naturalist John Nott in 1886. This rumor can also be traced all the way back to Aristotle, who developed an arcane theory about how bravery could be predicted by the size of an animal's heart. In his formulation, courage was viewed as proportional to the heat of the blood, which was in turn related to the size of the organ pumping it around the body: "When the heart is of large size the animal is timorous, while it is more courageous if the organ be smaller and of moderate bulk." Or so the grandfather of zoology said in his magisterial work *De Partibus Animalium* ("On the Parts of Animals"). He lumped the hyena, somewhat incongruously, in with "the hare, the deer, the mouse," along with "all other animals that are either manifestly timorous, or that betray their cowardice by their spitefulness" for having a disproportionately large heart. The persuasive details of Aristotle's theory may have been lost to time, but the idea that hyenas are cowards has persisted well into the modern age. Even the twentieth-century biologist's bible, E. P. Walker's *Mammals of the World*, published in the 1960s, stated authoritatively that spotted hyenas "are cowardly and will not fight if their prospective victim defends itself."

On safari one misty morning near Lake Nakuru, Kenya, I caught up with a pack of spotted hyenas hunting a zebra, their favorite prey. I'll admit it was a tough watch. When I arrived on the scene, the hyenas had already ripped the hide from the right flank of the zebra, exposing the beast's innards in the manner of a living, breathing Gunther von Hagens anatomical sculpture, as the skin trailed behind the animal like a piece of half-discarded clothing. The hyenas were now following this semi-eviscerated beast, waiting for its inevitable collapse. It was hard not to anthropomorphize the actors before me: the zebra seemingly dignified in the face of death, the hyenas cruel and cowardly. But survival is an unsentimental sport, and the hyena's hunting strategy is based on endurance. It often involves "testing" prey to see how much fight they have left, which could be interpreted as timidity but is better understood as a key part of their winning long game. There's no point risking fatal injury from a kick or a claw when all you need is patience.

So the idea that hyenas spend their time sneaking about and stealing the spoils from more "noble" animals like the lion is another misconception. Field studies have found that lions actually steal more kills from spotted hyenas than vice versa. The animosity between the animals is, however, very real: these two species are archenemies, locked in battle over territory and food. While lions may have the size advantage, the hyena makes up for this with intelligence. "Lions are not the sharpest tools in the shed," as Kay Holekamp put it. The hyena is a pathetic fool in *The Lion King,* but these feminist freaks are the brains of the savannah—and smarter than your average carnivore.

A FEW YEARS AGO, I got to spend a couple of days watching spotted hyenas in the Maasai Mara with Dr. Sarah Benson-Amram, an expert in hyena intelligence. "I think their reputation for being stupid has a lot to do with their gait," she told me. "They have this sort of lope which is very beneficial in terms of being superefficient runners, energywise; they can run really long distances. But it gives them this sort of awkward, dopey appearance."

To find the truth, Dr. Benson-Amram invented the world's first carnivore IQ test—a metal puzzle box with a meaty treat trapped inside that can only be released by using brains, not brawn. She has plonked her puzzle box before various predators from polar bears to panthers in order to gauge their problem-solving skills. She discovered that the animals that performed well tended to share a rich social life. Sarah believes sociability could be the evolutionary force responsible for the hyena's superior intellect.

Spotted hyenas gather in social groups bigger than any other carnivore. Their packs can number up to 130 individuals and they have been observed defending territories of up to 620 square miles. They live by the clan, and everything they do is tied up with the female dominance hierarchy that underpins it, but they don't remain together all the time. Instead, they spend much of their time in smaller splinter groups that coalesce in order to fight, hunt or feed. This arrangement is known as a *fission-fusion society,* and maintaining it demands sophisticated communication skills.

The aristocratic French naturalist Georges-Louis Leclerc, Comte de Buffon, dismissed the hyena's calls as simply sounding like "the sobs or reachings of a man in a violent fit of vomiting." But spotted hyenas have one of the richest vocal repertoires of any terrestrial mammal, including primates. They make a wide range of sounds including that famous giggle (actually a sign of submission), but the whoop—a quintessential sound of the savannah—is their signature call. It is a

ghostly echo that can carry on the wind for up to three miles, transporting with it a wealth of information about the caller, including its
identity, sex and age. Hyenas' big brains have evolved to memorize the
identity and rank of each of their clanmates. And they appear to be
able to recall each member's voice and status throughout their lives—
no mean cognitive feat, and one that ensures they have the political
savvy to both recognize friend from foe in a single call and negotiate
their strict social hierarchy, without endless conflict.

Sarah demonstrated this to me near a communal den where a half
a dozen females and their cubs were wiling away the heat of the day.
The adults were mostly dozing in the shade of an acacia tree while
their cubs tumbled about looking surprisingly cute. Sarah pressed play
on her iPhone and the whoop of a stranger hyena, not from this clan,
rang out from a portable speaker. Even though the recorded call was
slightly distorted, it still gave me goose bumps, stirring some atavistic
fear locked deep inside my limbic brain. A pack of hyenas would make
mincemeat of us in minutes. Impersonating a rival gang member was
asking for trouble.

Sarah wasn't taking any chances; we conducted our research from
the secure confines of a large Land Rover, positioned about a hundred
yards from the resting clan. Sure enough, as soon as they heard Sarah's
recording, the hyenas' ears pricked up and they looked over in our
direction, instantly on alert. They stood up and sniffed into the breeze
seeking more information. Their sense of smell is a thousand times
stronger than ours, with each clan throwing off a distinctive scent, a
fragrant flag they wave in the wind. One particularly well-built hyena
started to gallop towards us and whoop. My heart rate quickened. But
the hyena loped straight past our vehicle as if it didn't even see us. It
was searching for something that looked and smelled like a hyena, and
my sweaty head popping out of the top of a safari vehicle did not fit
the bill.

Sarah has shown that hyenas respond differently depending on
whether they hear the whoops of one, two or three individuals. This
means the spotted hyena sisterhood can, in some sense, count—a useful tool when trying to gauge whether to fight a rival gang. She has

also shown that rival hyena clans will use their numeracy and communication skills to band together and fight off a common enemy, such as the lion.

Despite being wired for aggression, spotted hyenas use their intelligence to keep the peace and collaborate. "Hyenas are very cooperative with their clan members and close relatives," she explained. "You see these sisters and they spend a lot of time together, eating, hunting and resting together, and they have long-term, very close relationships . . . While they can be very competitive, in many respects they are also very cooperative."

Ultimately, the hyena's extraordinary success at bringing down large prey, intimidating lions and raising their cubs in a hostile environment depends on their capacity for teamwork. Recent fieldwork suggests that the social structures of the spotted hyena are every bit as complex as those of baboons, and CT scans have confirmed that hyena brains have evolved in a similar frontal direction to primates, with the region involved in complex decision-making enlarged. They have even outperformed chimpanzees on certain cooperative problem-solving tests. This supports the idea that living in a complex fission-fusion society— as both chimps and hyenas, along with dolphins, other apes and, of course, humans, do—is key to the evolution of big brains. This may even help explain why our species evolved a brain that is seven times bigger than what would be predicted for an animal our size.

This shared portion of the evolutionary path may also provide the ultimate clue to our unswerving contempt for these calculating creatures. Humans and hyenas are long-standing enemies. The Australian anthropologist Marcus Baynes-Rock, who has spent several years living in Ethiopia studying the relationship between our two species, has some insights as to why.

Both humans and hyenas are highly intelligent social predators with origins on the African savannah, he explained to me. But the hyenas were there first, so when our distant hominin relatives came down from the trees, they were muscling in on the spotted hyena's patch. "There must have been a huge amount of antagonism," Baynes-Rock said. "Just looking at the way that hyenas and lions get along today—they absolutely hate each other—we can just imagine humans and hyenas at the beginning were the same—hating each others' guts." The early hominins were also at risk of being eaten by hyenas. "This is a slow, fatty, very edible primate, and the only way they have to protect themselves is forming large groups. Hyenas would definitely have taken the opportunity to eat a *Homo habilis* if it strayed too far from the group."

Baynes-Rock thinks the hyena's bone-crunching bite could even be the reason for the sparse evidence of early human evolution. "Most hominid remains are just teeth and jawbones. When you are finding teeth, it's almost a guarantee that the dead person has been through the digestive tract of a hyena, because that's all that comes out."

Our early ancestors had only very basic stone tools, and were probably scavenging more than they were hunting. They would not have been able to fight off a pack of hungry hyenas to protect their prize meal; a theory borne out by bones from the period that show cut marks from early stone tools mixed with tooth marks of hyenas, suggesting that hyenas were laughing at us *and* stealing our dinner for as much as 2.5 million years. No wonder we don't like them.

CHAPTER 5

VULTURE

Order Accipitriformes

The eagle attacks his enemies or his victims one on one . . .
Vultures on the other hand, join together in troops, like cowardly
assassins, and would rather be robbers than warriors,
birds of carnage rather than birds of prey . . .
—Comte de Buffon, *Histoire Naturelle*, 1793

Scavengers in general have a tough enough job commanding respect from mankind, but looking like the grim reaper *and* dining on the dead has done little to promote an unbiased understanding of vultures. The reputation of these magnificent raptors has long been haunted by a queasy combination of disgust and mistrust. The grandiose naturalist Georges-Louis Leclerc, Comte de Buffon was especially bombastic when it came to describing the vulture. "They are voracious, cowardly, disgusting, odious and, as with wolves, just as noxious during their lives as they are useless after their death," he said, throwing his thesaurus at the vulture in a frenzy of unflattering adjectives.

Our uneasy relationship with death, with which these birds seem to be so at one, rubbed off on their hunched shoulders. Early Christian taboos against touching corpses put vultures in a singular category of grotesquery. The Old Testament stigmatized them as unclean, "an abomination among the birds." They were seen as otherworldly creatures in possession of mystical powers. "Vultures are accustomed to foretell the death of men by certain signs," one twelfth-century bestiary warned. "Whenever two lines of battle are drawn against each other in lamentable war . . . the birds follow in a long column [and] show by the length of this column how many soldiers are to die in the struggle." These clairvoyant powers were, according to the chronicler, self-serving: "They show in fact how many men are destined to be the booty of the vultures themselves."

Vultures are unsentimental about what—or who—they eat. Medieval European battlegrounds, dotted with dozens of felled humans and horses, would have been the equivalent of an all-you-can-eat buffet. Battles often waged for weeks, frequently during the summer months—conveniently coinciding with the vultures' nesting season on the continent. Birds must have set up camp in Hitchcockian numbers, dispassionately picking over the dead to feed themselves and their chicks. An observer some distance away might thus be able to predict something about conditions on the battlefield by counting the vultures circling above.

Despite popular mythology, however, vultures do not stalk their prey while it's still living, and they are not able to predict mortality.

How vultures know where to congregate is indeed mysterious. Their eerie ability to arrive apparently out of nowhere, and in great numbers, at the scene of death has engendered a longstanding belief that their senses must be of supernatural strength. But precisely which senses are being used has been the subject of one of the longest and bitterest arguments in ornithology.

The defining feature of the vulture's dinner is that it stinks to high heaven. So it was long held that this bird's sense of smell must be the source of its mystical power of tracking down the dead. As the Franciscan monk Bartholomew the Englishman wrote in his influential thirteenth-century bestiary, "In this bird the wit of smelling is best. And therefore by smelling he savoureth carrions that be far from him, that is beyond the sea, and ayenward."

The scavenger's admirable olfactory ability was widely accepted and reported in later natural history books. Oliver Goldsmith, for example, grudgingly conceded in his *History of the Earth* that although the vulture's nature may be "cruel, unclean and indolent," their "sense of smelling, however, is amazingly great." He qualified this alleged talent with physiological evidence: "Nature, for this purpose, has given them two large apertures or nostrils without, and an extensive olfactory membrane within."

Just a decade later, the vulture was spectacularly robbed of its sense of smell by an ambitious American naturalist named John James Audubon. Today, Audubon is one of the most recognized names in ornithology, famous for his exquisite lifelike drawings of birds. But back in the 1820s, he was a wandering go-getter, hawking his paintings around Europe and hungry for a bit of fame. He got what he wanted by courting controversy at a gathering of the hallowed Edinburgh Natural History Society in 1826.

The rambling yet provocative title of his subsequent paper betrays Audubon's desire to make a big splash with the stuffy Victorian natural history crowd: "An Account of the Habits of the Turkey Buzzard (*Cathartes aura*) Particularly with a View to Exploding the Opinion Generally Entertained of Its Extraordinary Power of Smelling."

In his audacious paper, Audubon explained how, as a child growing up in France in the late 1700s, he had been taught that vultures scavenged using their sense of smell. This didn't make any sense to the budding ornithologist, who labored under the spartan belief that "nature, although wonderfully bountiful, had not granted more to one individual than was necessary and that no one was possessed of any two of the senses in any a very high state of perfection; that if it had a good scent, it need not the acuteness of sight." Years later, while living in America, Audubon had begun testing his theory by casually creeping up on wild turkey buzzards (also commonly known as turkey vultures), but they seemed only to be spooked by his sudden appearance, not his body odor. He decided to take the matter further and "assiduously engaged in a series of experiments to prove, to myself at least, how far this acuteness of smell existed, or if it existed at all."

Audubon's grand experiments basically amounted to a very smelly game of hide-and-seek involving some dead animals and a committee of wild vultures. First, he roughly stuffed a deer skin with straw and

Audubon's illustration of turkey vultures in this famous avian encyclopedia, *Birds of America* (1827–1838), is exceedingly true to life. Which makes it all the more unfortunate that the great ornithologist got the species muddled up with black vultures, a mistake that sparked one of the biggest controversies in ornithology.

left it in a meadow with its feet in the air. His taxidermy skills were not the finest, but the misshapen beast still quickly attracted a vulture, which launched a futile attack on the dummy's clay eyes, then "voided itself freely" and unpicked some stitches from the dead deer's rear end, releasing "much fodder and hay." The disillusioned bird then flew off and killed a small garter snake as comestible compensation, thereby proving to Audubon's satisfaction that the bird had used vision to hunt.

Next, on a hot July day, Audubon turned his sights to the vulture's sense of smell. He hauled an exceedingly smelly, putrefying pig into some woods, which he then hid in a ravine, obscuring the animal from plain view. Although vultures cruised overhead, none came down to seek out the source of the stench. Audubon's suspicions were once again confirmed: these vultures did not scavenge by smell.

Audubon's results were far from conclusive, but the man was a consummate showman with a reputation on an upwards trajectory after the phenomenal commercial success of his book *Birds of America*. Many respected members of the ornithological establishment stampeded to support his explosive proposition. There was, however, one very vociferous exception from the herd: an aristocratic adventurer by the name of Charles Waterton.

To describe Waterton as an eccentric would be to contain him within a crowded category that he alone should own. The Squire's outrageous accounts of his exotic escapades, which allegedly included riding a crocodile and punching a boa constrictor in the face, had earned him a certain infamy. At home in Walton Hall, his actions were no less unconventional. He was said to have a habit of hiding under the table at dinner parties to bite his guests' legs like a dog, and delighted in elaborate, taxidermy-based practical jokes. A particularly inspired prank involved his fashioning an effigy of one of his (many) enemies out of a howler monkey's buttocks.

Despite such antics, Waterton was a talented naturalist—an original thinker whose sideways view of the world afforded him a relatively unprejudiced understanding of nature. He was, for example, one of the first to stand up for the sloth, whose "extraordinary formation and singular habits" he felt were in fact reason for us "to admire the wonderful works of the Omnipotence."

Waterton had written about the turkey vulture's unique talent for smell in his own bestseller, *Wanderings in South America*, and took the cocky American's counterclaims as a personal attack on his admittedly somewhat shaky credibility. This spurred the Squire to wage a lengthy war of withering and occasionally witty words against Audubon. These missives were printed in the *Magazine of Natural History*—the nineteenth-century equivalent of a very public Twitter spat.

"I grieve from my heart that the vulture's nose has received such a tremendous blow, because the world at large will sustain a great loss by this sudden and unexpected attack upon it," wrote Waterton. "Moreover I have a kind of fellow-feeling, if I may say so for this noble bird." As the self-appointed leader of what became known in ornithological circles as "the Nosarians," the Squire offered to "carefully gather up the shattered olfactory parts, and do my best to restore them to their original shape and beautiful proportions."

Waterton took a surgeon's knife to Audubon's experimental acumen, his reputation and even his scientific prose. "Its grammar is bad; its composition poor; and its statements are so unsatisfactory," he complained. "In my opinion, any person who reads the paper with any moderate share of attention will feel inclined to condemn it to the same kind of fate as that to which the curate and the barber condemned the greater part of Don Quixote's library." He called Audubon a liar and a charlatan, entirely incapable of having penned a bestselling bird book. "Mr. Audubon's story of a rattlesnake swallowing a large American squirrel, tail foremost, still sticks in my throat."

Audubon remained aloof throughout, preserving his dignity with silence. Instead, he encouraged his growing band of "Anti-Nosarians" to answer Waterton's attacks on his behalf. Chief among his defenders was a Lutheran pastor, the Reverend John Bachman, who attempted to settle the quarrel by repeating Audubon's experiments, augmenting them with a few of his own, in front of a select committee of learned gentlemen in his hometown of Charleston, South Carolina.

The pastor's experiments were brutal and bizarre in equal measure. In one, he commissioned an oil painting of "a sheep skinned and cut open," which he then placed in his garden, about three yards from a hidden pile of decaying offal. The painting was crude—not up to Audubon's fine standards—yet the undiscerning birds still attacked the artwork with great gusto. Bachman said the vultures "seemed much disappointed and surprised" at their lack of satisfaction, but not once did they move towards the meat festering nearby. He repeated the escapade fifty times, which, according to his report, "proved very amusing" to the learned men in attendance.

In a bonus experiment, he enlisted some "Medical gentlemen" to blind a vulture in order to test a rumor doing the rounds that the bird could repair its perforated peepers by simply placing its head under one wing. When the sorry bird failed to restore its own eyesight, Bachman saw his opportunity to probe its remaining senses. He consigned the wounded vulture to an "out-house" containing a dead hare and watched to see if the poor blind bird could sniff it out. It could not. In an uncharacteristic fit of charity, the reverend conceded that "the bird might not have been wholly free from the pain inflicted by the operation." Indeed.

Beyond that, Bachman showed no sign of remorse; his only concern was whether his malodorous experiments might "become offensive to the neighbours." This weighed on him mightily, so he called an end to the investigation, satisfied that he had settled the score. Before publishing his findings, he took the rather heavy-handed measure of pressing his medical men into signing a contract confirming they'd witnessed definitive proof that vultures scavenge "through their sense of sight and not of smell"—proving that the pastor's campaigning techniques were as punishing and peculiar as his quest for scientific truth.

Waterton's response to Bachman was suitably scornful. "Pitiable, indeed, is the lot of the American vulture! His nose is declared useless in procuring food, at the same time that his eyesight is proved to be lamentably defective," he grumbled. "I am now quite prepared to receive accounts from Charleston of vultures attacking every shoulder-of-mutton sign in the streets, or attempting to gobble down the painted sausages over the shop doors, or tugging with might and main at the dim and faded eyes in some decaying portrait of the immortal Doctor Franklin."

It did not pay to annoy "the Squire." The government jobsworth who
dared to tax Charles Waterton on the exotic specimens he collected abroad
found his likeness fashioned from a howler monkey's buttocks and labeled
"A Nondescript." The taxidermist's prank was further immortalized in this
copper engraving from his book *Wanderings in South America* (1825).

Over the course of five years, Waterton wrote no less than nineteen letters to the *Magazine of Natural History* attacking Audubon and
anyone in his orbit. When the journal finally stopped publishing his
letters, he reportedly continued to print and distribute them himself.
His efforts were futile. His impenetrable, rambling diatribes, punctuated with sardonic ad hominems and obscure Latin phrases, won him
few allies. Audubon's anti-Nosarians branded the Squire "stark, staring
mad" and resolutely refused to alter their opinion of the flashy American. The louder Waterton shouted, the more he was ignored. In the
end, he was forced to give up.

Which is a shame, because he was right.

IT WOULD TAKE SCIENCE almost 150 years to catch up with Charles Waterton. Meanwhile, the waves from Audubon's explosion engulfed an expanding circle of anatomists, naturalists and ornithologists, who went to work performing ever more incredible experiments on a widening variety of birds.

In one of the more ludicrous, a domesticated turkey was substituted for a turkey vulture, its food hidden in a saucer containing sulfuric acid and potassium cyanide. The turkey did not survive the plumes of poisonous gases long enough to reveal whether it could smell its supper or not.

The vulture's other senses, both real and imaginary, were also drawn into the debate. In the early twentieth century, a chap called P. J. Darlington argued the bird actually used its ears, listening for the distant buzz of hundreds of flies to find its meal. Another theorist, named Herbert Beck, reverted to medieval mystical thinking in a 1920 paper entitled "The Occult Senses in Birds," which supposed that the vulture was in possession of a mysterious "food-finding sense" that remains completely inexplicable to humans because we do not possess it.

The turkey vulture's sense of smell was finally restored in 1964, when Kenneth Stager, another maverick American field scientist, presented conclusive evidence based upon his years of careful and clever experimentation, along with a little serendipity. Stager's major revelation came during a random exchange with an employee of Union Oil who let slip that since the 1930s the company had been exploiting the turkey vulture's keen sense of smell by using the birds to locate leaks in their gas pipelines. They had started adding ethyl mercaptan, the rotten cabbage smell that flavors stink bombs and flatulence, to their natural gas, as it invariably attracted vultures, which then homed in on and exposed any leaks before humans could detect them. Stager realized the same compound is also released by decaying corpses. And sure enough, when he used a mercaptan-dispensing machine to fart rotten odors into the Californian sky, the turkey vultures soon came circling.

The decades of confusion over the vulture's sense of smell can be traced to a handful of basic misunderstandings. First, it would seem that Audubon wasn't quite as attentive to his birds as one might suppose. Some of the species he described showed an interest in hunting down live animals as well as dead, which suggests they were black vultures (*Coragyps atratus*) and not, as he identified them, the similarly black-plumaged turkey vultures (*Cathartes aura*).

Second, it was assumed that all vultures would exhibit similar olfactory abilities. They don't. The twenty-three species of vultures fall into two distinct groups—Old World vultures, which inhabit Africa, Asia and Europe; and New World vultures, which live in the Americas. Though they look and behave in a similar fashion, the two groups are only distantly related, not even falling into the same family, let alone genus, of the animal kingdom. It turns out that all vultures hunt by sight but only a handful of the New World vultures, including the turkey vulture, hunt by smell. Crucially, black vultures are among those that do not use smell.

Third—and despite conventional wisdom—vultures are surprisingly discerning about what they eat. Like us, they prefer to eat dead herbivores rather than carnivores, and they don't like them *too* rotten. Yet another rather eccentric experiment, this one conducted by David Houston in the 1980s, involved hiding seventy-four chickens in a Panamanian jungle. It revealed the turkey vulture's perfection of putrefaction was a perfectly *al dente* two days after death, neither older nor younger. American turkey vultures may well have turned up their nose at Audubon's perished pig and Bachman's ancient offal simply because the meat was too far gone.

In more recent years, news of the turkey vulture's olfactory detecting talents has attracted the interest of the German State Office of Criminal Investigation, which has pioneered a scheme to train vultures to replace sniffer dogs. Police officer Rainer Herrmann got the idiosyncratic idea after watching a wildlife documentary that bragged of the birds' smelling skills. He hoped the vultures, fitted with GPS trackers and followed by a fleet of land cruisers, would be able to cover more ground, more quickly, than dogs.

A turkey vulture from a local bird park was chosen for the pilot and given the predictable nickname of Sherlock as well as a dedicated

German Alonso attempts to encourage trainee detective Sherlock in sniffing out some missing persons (but hopefully not in attacking their eyes or anus).

personal trainer with the very German name of German Alonso. With such a whimsical cast, the scheme got plenty of media attention, and before too long requests for the vulture's services were flooding in from forty police departments around the country.

Alonso had some reservations about his charge. He thought vultures might struggle with differentiating between dead humans and dead animals, causing a fair few false alarms. But he showed surprisingly little concern about the prospect of the birds eating any evidence they found. "That will happen and you can't stop it," the trainer announced to the national newspaper, before adding, "But they won't remove the entire corpse, they can't eat that much. And if they take a nibble, what the hell, the victim will be beyond help anyway"—an attitude unlikely to soothe the nerves of a missing person's mom, or please a forensic scientist in need of pristine clues.

Sherlock, however, was significantly less enthusiastic about his new job than everyone else. He didn't like to fly when searching for his training material—a mortuary cloth previously used to wrap a dead body. Instead, he hopped around the ground nervously investigating a small area on foot. At times he was so anxious he would hide in the woods or bolt when given the command to search. Two younger vultures, named Miss Marple and Columbo, were brought on board to make him feel part of a big vulture detective family. All they did was fight.

Wild turkey vultures may be good at spotting gas leaks, but a handful of captive vultures cannot, despite their pet names, be relied upon to solve crimes by following their nose. It shows that, even in modern times, the vulture's sensory skills are subject to exaggeration. Police officer Herrmann was under the impression the birds would be able to sniff out a dead mouse from more than a thousand yards. This is unlikely, as recent research has shown that the birds must fly low in order to pick up the scent of the dead. So while turkey vultures do scavenge by smell—exactly as Waterton claimed—their olfactory abilities are nowhere near as good as a sniffer dog's, and possibly not that much better than a human's.

The vulture's eyesight has also been elevated to mythological status. In southern Africa, it's believed the bird's ocular abilities are so keen they possess what is called "clear sight"—the ability to see into the future.

A few years ago, I visited the main *muti*, or traditional medicine market, in Johannesburg to investigate. Among a sea of stalls selling dismembered animal parts, I discovered dozens of people trading in small vials of vulture brains, which I was told were smoked or snorted to attain clairvoyance. Since the arrival of the Lotto, vulture brains had become the biggest seller in the market—something vulture conservationists could never have predicted, no matter how much of the stuff they inhaled.

Marginally more credible is the bird's oft-quoted ability to spot a carcass from 2.5 miles away, a mainstay of most vulture "fact sites" online. Anatomists dissecting vultures' eyes have been surprised to discover the bird's eyesight may only be twice as good as our own. Vultures lack binocular vision and, thanks to a well-developed ocular ridge that protects their eyes from the glare of the sun (and which bestows the birds with their characteristic ferocious stare), they also have significant blind spots.

Vultures' spooky ability to arrive at the scene of death in such great numbers so quickly is down to the assistance of another highly developed organ: their brain. Vultures are canny creatures that watch and learn from each other. In most instances, a vulture won't discover a carcass itself; instead, it spots a column of circling birds, which *are* visible from several miles away, and heads in their direction. Young vultures spend a significant amount of time learning scavenging techniques from their parents. And families of birds stay close to each other, with clans of related birds roosting together. Different vulture species also roost together in large numbers. Scientists have proposed that these scavenger social gatherings are a way for the birds to acquire information about the location of their ephemeral food sources.

ONE THING ABOUT VULTURES is for certain: their arrival is rarely welcome, even in our modern world. When, for example, a kettle of around five hundred turkey vultures, pushed north by global warming, decided to add Staunton, Virginia, to their winter migratory route in 2011, the human residents of this picture-postcard historic town were not exactly pleased.

"They're ugly as ****. I walk around the corner and fifty of them are sitting on gravestones, hissing. It's like living in a horror movie. If it were up to me, I'd kill every one of them."

"They're disgusting," a local woman told the *Washington Post*, describing her new neighbors. The birds carpeted her neatly manicured driveway with excreta as if it were a giant Jackson Pollack. That's to be expected: turkey vultures are prodigious pooping machines. They practice *urohidrosis*, a scientific euphemism for crapping on your legs to keep cool. It's not the most elegant of thermoregulatory solutions, but it is nevertheless a crafty substitute for sweating, which birds cannot do.

The results of urohidrosis aren't just offensive to the eye. "They smell like ammonia and sewage," another Virginia resident said.

Unfortunately for the bevultured citizens of Staunton, the birds are protected in the United States and cannot be killed without incurring serious fines. One frustrated local tried to improvise by shooting at the vultures with paint balls, only to discover the birds have a rather unsavory means of fighting back. "It vomited on my son," he told the *Washington Post.* "It was like a half pound of ground beef on his shoulder. It was so disgusting. We got it off him. Got his shirt off. And got him to stop screaming."

Yes, vultures' first-line defense, if full, is to chuck their dinner at you. Given that their supper probably wasn't terribly appealing first time around—the Virginia turkey vultures were likely dining on a menu of road kill and animal excrement—it's easy to see how defensive puking pushed neighborly relations over the edge in Staunton. (The US Department of Agriculture's "environmental police" were called in to run the avian outlaws out of town. Dead vultures were hung in roosting sites, firecrackers were set off and, in a dramatic final showdown, a number of birds were permanently dispatched from the planet.)

As the people of Virginia discovered, the vulture boasts an arsenal of unappealing habits that have long offended human senses. "The sloth, filth and the voraciousness of these birds almost exceed credibility," spat the big-talking Comte de Buffon in another of his tirades against the vulture. Even Charles Darwin couldn't stomach the bird's personal habits. In *The Voyage of the Beagle*, he described the turkey vulture as "a disgusting bird, with its bald scarlet head, formed to wallow in putridity." Not only was Darwin's description colored by prejudice, but it was also probably wrong. Other avian scavengers—including the giant petrel—do perfectly well getting gory with feathered heads. The vulture's baldness may instead help the bird stay cool, thereby joining urohidrosis as yet another aesthetically challenging adaptation for regulating body temperature.

LOOKS CAN BE DECEIVING. Just because these big scavenging birds are wandering around with legs encrusted in their own feces doesn't mean they are dirty. And scavenging, though wildly unappealing to posh Parisian naturalists, is far from a degraded means of procuring dinner.

Indeed, it's quite the opposite, as I learned when I spent some time with conservationist Kerri Wolter at her vulture sanctuary in South Africa, where she has been working with the birds for more than fourteen years.

Kerri is on a crusade to change the public perception of these much maligned creatures before it is too late. Globally, vultures are the fastest declining category of bird. Of the nine species found in southern Africa, all but one are threatened with extinction.

It's a short drive from the Johannesburg airport to the sanctuary, nestled somewhat incongruously at the edge of Pretoria's concrete sprawl. Kerri cares for around 130 rescued birds, victims of power-line electrocutions or accidental poisoning. Most of her guests are critically endangered Cape vulture, or *Gyps coprotheres*, an Old World species.

I arrived just in time for their lunch. Kerri put me to work straight away, helping her wheel a freshly killed cow carcass into the main enclosure, home to a breeding group of a dozen or so birds.

The first thing that struck me was their size. Weighing in at around twenty-two pounds and standing about three feet tall, Cape vultures are the largest Old World vulture in southern Africa and one of the world's bulkiest flying birds. Being big is important for a carnivore that must rely on scattered and infrequent food sources. If you can't go out and kill something when you are hungry, it helps to be able to gorge on food when it's available and live off your fat reserves. Size also helps intimidate fellow scavengers competing for morsels from a communally shared corpse. These birds certainly intimidated me. I was already feeling a certain amount of trepidation about entering their cage when Kerri advised me to put on sunglasses, just in case the vultures took a fancy to my eyes.

Bird scavenging, she explained, has developed into highly specialized behaviors, with vulture species divided into various "guilds" according to the design of their beaks. There are "tearers," "peckers" and "pullers," each type working in an uneasy, belligerent team to get the job done on a shared carcass. I guessed that the pullers were the ones most likely to be interested in my eyes, and I was right.

"In South Africa, the lappet-faced vultures are the knives," Kerri said. These tearers are able to rip through tough hides using their short necks and powerful narrow beaks, which can exert 1.4 tons per square centimeter of pressure. And Cape vultures are indeed pullers, with long muscular necks and sharp beaks ideal for reaching deep inside a carcass to gobble up soft flesh and organs.

Despite their imposing size, the Capes cannot tear open the carcass themselves. So if there are no lappet-faced vultures around, the only way for them to get to their meal is through a natural orifice—like the eyes or anus. "They go for the soft bits," Kerri explained. I double-checked my sunglasses and backed into the fence.

Kerri was proud of her wards. "Vultures are the most efficient scavengers. They have a special adaptation with which to clean meat off bones by their hooked tongues. Strong feet and legs to stand on and hold the carcasses down. Some with long bare necks to dig into the carcass and eat from the inside out." She noted that one species had even evolved to consume the bare bones, once the peckers and pullers have removed all the flesh.

This division of labor explains why you'll sometimes see vultures standing around, looking a bit gormless, near a freshly dead animal—a behavior that helped contribute to the myth that they prefer putrid flesh to fresh kill. "There'll be an elephant carcass that's been sitting there for days and the vultures haven't touched it," Kerri said. "The simple reason is not because they prefer rotten carcasses, it's that they can't get in, so they're waiting for it to become soft so they can rip it open."

Once the vultures get inside a corpse, the resulting frenzy can be quite a show. Buffon described feeding vultures as "displaying the bitterness of unprovoked rage." I thought it was more of a dark comedy, something directed by Quentin Tarantino and starring the original "angry birds" stuffing their faces with the bilious speed of an American hot dog–eating contest. There was much strutting and hissing, drooling and sneezing, posturing and pecking, as the cow quickly disappeared behind a writhing mass of spaghetti necks and bald heads caked in bloody remains and buzzing with flies.

Our revulsion over the vulture's eating habits is for good reason. If we were to feast on rotting flesh, we'd likely get very sick very quickly. Vultures survive by obliterating disease-bearing bacteria, including some very unforgiving ones, like botulism or anthrax, with some of the strongest stomach acids in the animal kingdom, with a pH similar to that of battery acid. As an added bonus it makes their excreta so caustic they can disinfect their feet after dinner simply by defecating. Kerri told me that vulture excrement is such an effective disinfectant that I could use it to clean my hands before we ate our own lunch. I decided to take her word for it.

Vultures also help prevent the spread of disease. By hoovering up pestilence and pooping out purity, vultures are their own highly effective, double-quick forensic cleanup unit. One hundred vultures can strip a carcass in twenty minutes, before contagion has had a chance to set in or spread very far. What little attaches itself to a bird gets washed away in a cleansing stream of ordure.

The speed with which vultures consume their food has been reviled by natural historians as "a revolting sight of selfish greed," but Kerri said it should be rebranded as an act of heroic generosity. Recent studies have shown that in areas where there are no vultures, carcasses take up to three to four times as long to decompose, a situation that favors contagion.

You can see the costs to humans in places where vultures have suffered dramatic declines. "Look at India and Pakistan, where the government has spent over thirty-four billion dollars on human health issues because vultures have been almost eradicated," Kerri said. India's vultures have been virtually destroyed by accidental poisoning after eating the carcasses of cattle treated with an anti-inflammatory drug called diclofenac; it's estimated that up to 99 percent of the three main species have been killed. Without so many vultures around, there's an excess of carrion. The secondary effect has been a massive increase in feral dogs and a huge rise in rabies.

"The amount of money that's going in for rhino poaching in comparison to vulture conservation is astronomical," Kerri said. "There's only a certain amount of funding for scavengers because people don't

like them, which is crazy because if we lose our rhinos, yes, it's going
to be sad. I love rhinos, but the world will continue. If we lose our vultures, there will be collapses throughout Africa, and it's going to affect
every single one of us."

To TRULY APPRECIATE a vulture's beauty, Kerri said, is to watch one
fly.

Vultures need to be able to cover vast distances, as economically
as possible, as they search far and wide for their next meal. Weighing
about the same as a human toddler doesn't exactly facilitate energy-
saving flight. Simply getting airborne is tricky enough, let alone traveling thousands of miles without the need for hours of exhaustive
wing-flapping. Vultures have evolved to crack this problem: they can
glide at speeds of up to fifty miles per hour while spending hardly any
energy at all.

To understand how vultures achieve this aerodynamic marvel I was
encouraged to jump off a mountain. Kerri and her partner, Walter,
a keen paraglider, assured me that this experience would bring me
closest to the magnificence of these birds. I thought it sounded like a
reasonable idea until I got to our launching point on the crest of the
Magaliesberg range, an ancient crust of almost sheer cliffs one hundred times older than Everest and nearly three thousand feet high.

The biggest vulture species, like the Cape vultures that nest in the
craggy contours of this aged escarpment, must make use of altitude
in order to launch themselves into the air with minimal exertion.
Their impressive eight-foot wingspan then hitches a lift on thermal
columns, rising updrafts of hot air, which propel them, like unseen
elevators, thousands of feet into the sky. Today, I would be doing
something similar, strapped to Walter and flying tandem under his
trusty glider.

I peered over the edge of the cliff to a valley shimmering far below
and felt decidedly unbirdlike. Paragliding is a counterintuitive sport.
We had to literally run off the cliff into thin air, the ultimate leap of
faith, followed by a terrifying moment of weightlessness. Then the
glider's wings caught the rising air and we started to circle upwards.

Initially, we rode the thermals alone. Then, as if out of nowhere, we were joined, one by one, by spiraling vultures. As Walter had predicted, the birds were very different in the air, curious and quite playful. And they *were* magnificent.

For us, the thermals were like an invisible roller coaster, bumpy and unpredictable, if not for the constant peeping of Walter's altimeter. We were jolted, tugged and tripped up by capricious currents. The vultures were apparently at one with the wind, their massive wings barely twitching as they effortlessly swooped in close to give us a good eye-balling, before peeling off and shooting high above us, as if flaunting their aero-perfection.

THE VULTURE'S WINGS are so perfectly adapted for soaring flight that they provided Wilbur Wright with a model for stabilizing the wings of his first successful aircraft, which he developed after hours of observing turkey vultures. Unfortunately, today the conceptual descendants of the Wright brothers' first flight are colliding with their inspiration in our overcrowded skies.

Bird strike, as it's known in the aviation business, is a hazardous affair, one that costs the US government more than $900 million a year and has destroyed thirty aircraft since 1985, making birds significantly more lethal than terrorists. In 2009, a flock of migrating Canada geese famously forced an Airbus to land in the Hudson, after they flew into the plane's engines. This might explain why the American military has someone testing their latest jet-engine aircraft for bird resistance by firing chickens out of cannons at point-blank range. Dead chickens, one presumes.

In an attempt to better understand the "enemy" and predict its moves, the Feather Identification Lab of the Smithsonian Institution in Washington, DC, takes forensic ornithology quite seriously. There, researchers perform DNA sequencing to ascertain the most common bird threats to planes. Every week the institute receives hundreds of packets of "snarge"—the bloody remains of when bird meets plane. Turkey vultures win the prize for No.1 most damaging bird to American aircraft.

The most puzzling snarge cases could belong to an episode of
The X-Files. A Smithsonian snarge detective named Carla Dove told
Wired magazine, "We've had frogs, turtles, snakes. We had a cat once
that was struck at some high altitude." After a certain amount of
head-scratching about how such typically terrestrial creatures could
have been caught more than a half mile up in the air, the researchers
realized they must be quarry dropped during flight by loose-taloned
birds of prey. Thus, Buffon's magnanimous eagles are equally to blame
for bringing down American planes, but for butter-fingered reasons.

Over the years the Smithsonian's elite snarge force has identified
about five hundred species of bird and forty species of terrestrial mammal, including a rabbit that hit a flight at over fifteen hundred feet. But
perhaps the most incredible was a collision that forced a commercial
craft to make an emergency landing on the Ivory Coast after an impact
at over thirty-six thousand feet, that's seven miles up in the air—the
highest known flight of a bird ever recorded. The feathered remains
were identified, with a fair amount of fanfare, as a Rüppell's griffon
vulture. The bird earned its crown (and messy end) through the assistance of a specially adapted form of hemoglobin that enables its blood
to absorb oxygen at pressures so low most other animals would pass
out.

That griffon vulture was probably a high-flying freak, but vultures
have been known to cruise at more conservative, but nevertheless
head-spinning, heights of over nineteen thousand feet, or nearly four
miles. By soaring from thermal to thermal, they're able to chase the
dry season as it creeps across the African continent and lays waste to
infirm animals.

One recent study observed a Rüppell's griffon vulture traveling
north from its nest in Tanzania, across Kenya, to a region in Sudan and
Ethiopia as it searched for food. Such border-hopping has occasionally
gotten these birds, and their human saviors, in political hot water. A
long-standing project to reintroduce endangered griffon vultures back
to their native Israel has been thwarted by paranoia and politics. Researchers at Tel Aviv University have begun tagging and tracking vultures, but the birds' long-distance lifestyle means they frequently fly

over the country's borders. Such is the fraught state of international relations in the Middle East that these vultures have been captured and accused of being Mossad spies. In 2011, the Saudi Arabian government arrested a bird they believed was part of a Zionist plot. Three years later, another vulture was detained in Sudan. And in 2016, the United Nations was forced to intervene and return a vulture apprehended by Lebanese villagers who were suspicious that the bird's GPS tracker was in fact a covert camera.

Ohad Hatzofe, the exasperated conservationist behind the scheme, pointed out that it's not terribly surreptitious, labeling a spy's legs with your email address. His somewhat sarcastic response to a Middle Eastern newspaper that asked him about the project was that if he was indeed recruiting secret agents, he might choose ones "less interested in dead camels and goats." It seems humans just can't help being mistrustful of vultures.

CHAPTER 6

BAT

Order Chiroptera

An animal, like the bat, which is half a quadruped
and half a bird, and which, upon the whole, is neither
the one nor the other, must be a monstrous being.
—Comte de Buffon, *Histoire Naturelle*, 1749–1788

The world's only flying mammal has become a rather unlikely You-Tube star. Not because of its cute furry face and toothy grin, but for something more menacing—its uninvited presence in people's homes. There's an entire subgenre of man-versus-winged home invader videos online, generally featuring an out of his depth dad in a desperate bid to defend his frantic family. Hysterical moms cover their hair and crawl along the floor sobbing, while babies get locked in bathrooms and everyone screams like doomed teenagers in a slasher flick.

And yet the terrifying monster in these real-life domestic horrors is little more than a tiny disoriented insectivore.

Bats have probably been flying into human homes ever since we've been building them (for that matter, ever since we first sought shelter in a cozy cave—which strictly speaking makes us the invaders of *their* home). The bats are simply searching for somewhere to roost, or chasing insects; they aren't interested in us, which does nothing to assuage our fear.

Today, a person with a persistent fear of bats has a bona fide syndrome. It's called chiroptophobia, after *Chiroptera*, Latin for "winged hand," the defining characteristic of the taxonomic order that encompasses the eleven hundred or so known bat species. Alas, a common therapy offered by therapists in America for this irrational fear and loathing is controlled exposure to a room full of bats, a nightmare in just about anyone's book, phobia or not.

A recent survey by a bat conservation group discovered that one in five perfectly sane British people also claim to hate bats. A common perception is that they are flying vermin, or, as the bat-hating comedian Louis C.K. called them, "disgusting" and a "rat with leather wings." The surveyed Brits believed that bats are blind, malevolent creatures intent on getting stuck in their hair, sucking their blood and giving them rabies—most of which is total rubbish.

Bats are actually more closely related to humans than rodents; they can see perfectly well, thank you very much (some fruit bats even have color vision that's three times as good as our own); and they have an acutely tuned echolocation system that prevents them from flying anywhere near even the most bouffant 'do. Finally, only three species have vampiric tendencies, and you are more likely to get rabies from a dog or a raccoon (less than 0.05 percent of bats carry the disease).

Bats are in fact more Buddha than Beelzebub. They are among the most magnanimous neighbors, charitable friends and generous lovers in the animal kingdom. They save us billions of dollars every year by eating insects that cause devastating diseases and destroy our crops. They are also the key pollinators for many tropical flowers, including bananas, avocados and agave. Without bats, there would be no tequila (which may, or may not, be a good thing for humanity). Frankly, they are more man's best friend than the dog.

PUBLIC RELATIONS HAVE NEVER run smoothly for the bat. Along with the vulture, it is one of the few animals that the Bible lists as unclean, which is a little harsh for an animal that can spend up to a fifth of its time grooming and is probably significantly cleaner than most of the big book's holy scribes were.

One early Roman author, Divus Basilius, took it a step further: "The nature of the bat is blood-related to that of the devil." Their physicality is partly to blame for the opprobrium. The arrangement of their bodies, limbs and faces, complete with forward-facing eyes and toothy smiles, are unnervingly human but decidedly inhuman at the same time. Once artists started painting images of bat-winged figures to represent the Devil—like those in the famous representation of Dante's *Inferno*—the negative branding was complete.

The natural history books of the medieval period were penned by religious types who viewed the bat's neither-here-nor-there nature with grave suspicion. The British clergyman and naturalist Edward Topsell felt obliged to include these deviant winged creatures in his

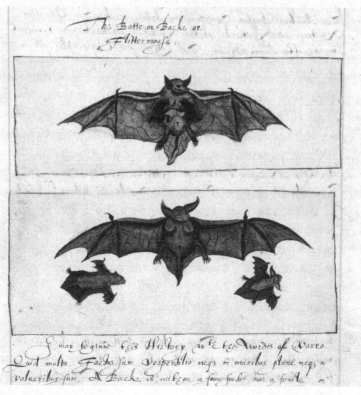

The bats from Topsell's *Fowles of Heaven* (c. 1613) have a distinctly diabolical look about them as they vaunt their unbirdlike breasts and gnashers.

seventeenth-century book of birds, *The Fowles of Heaven*. The bats' refusal to conform to the model of their angelic avian cousins caused the man of the cloth much consternation. He was particularly bothered by the tiny mammals' distinctly unbirdlike breasts, teeth and love of darkness, which encouraged all manner of satanic comparisons. Topsell illustrated his concerns with sketches of bizarre bosomy beasts wearing wide, wicked grins. In a final, hallucinatory postscript, he accused these errant birds of guzzling the oil out of his church lamps.

As science headed into the Enlightenment, biologists continued to be bothered by the bat's anomalous anatomy, which defied classification. The imperious Georges-Louis Leclerc, Comte de Buffon, gave the bat a suitably damning write-up in his encyclopedia. "It is an imperfect quadruped, and a still more imperfect bird. A quadruped should have four feet, and a bird should have feathers and wings." Buffon was also offended by the bat's nether regions, which appeared to have been borrowed from another species, perhaps—rather alarmingly—his own. "The penis, which is pendulous and loose," he wrote, "is a thing peculiar to man, [and] the monkeys."

I, too, was quite taken aback the first time I got up close and personal with a bat's penis. It was about ten years ago, in a remote corner of the Peruvian Amazon where I had joined Dr. Adrian Tejedor, a Cuban Chiroptera specialist, mist-netting for bats. This involved setting up what looked like a very large, very fine badminton net in an opening in the rainforest, then waiting, like spiders, for some bats to fly into our trap (the netting so fine they could not detect it). We sat for several hours in the dark, sticky gloom of the night, our flashlights off so as not to scare the bats.

Our first guest was the lesser spear-nosed bat. Tejedor was excited; he had not seen this species for nine years. I was excited by the size of its penis, which dangled almost down to the bat's knee. It seemed anything but aerodynamic. "Penis length in mammals appears to be correlated with the degree of promiscuity in females," Dr. Tejedor said. Lady lesser spear-nosed bats must be absolute strumpets, then, since the males are famous for their especially elongated appendages.

Could that grin be any wider? Me and my big penis bat prize, from a night mist-netting in the Peruvian Amazon (the bat looking rather less thrilled with the unwarranted exposure).

The longer their penis, the farther they can insert their sperm into the female, giving their seed a head start over that of their romantic rivals.

Since my first chiropteran date had been with the Dirk Diggler of bats, I thought I might have a warped opinion of their wedding tackle. But the work of another bat specialist, Dr. Kate Jones, professor of ecology and biodiversity at University College London, has revealed that some bats have not only pendulous penises but big, bulging testicles to boot. In the war of the sexes, upping sperm production can be another useful tactic for males trying to outsmart highly promiscuous females—especially those, like many bats, that have the sneaky ability to store a male's sperm in their reproductive tract (to ensure fertilization happens at the most optimum time).

Dr. Jones knows a thing or two about bat genitalia. She was part of a team who methodically measured the gonads and brains of 344 bat species. As warm-blooded animals that fly for a living, bats exist on an energy knife-edge, and both these organs are metabolically expensive. Jones's team had a hunch that there was a trade-off between the two organs, and there was: monogamous bat species had tiny testes and big brains, while promiscuous ones had quite the reverse. One particular swinger sports a pair of testicles that accounts for a whopping 8.4 percent of its body weight—the Rafinesque's big-eared bat (artfully named after its *other* enormous organ) flies about hauling the human equivalent of a pair of large pumpkins between its thighs, guided by an inversely proportional, dumbed-down mental capacity.

The degree of male promiscuity is not a driving force in the brains-to-balls trade-off. Instead, it is the *female's* philandering that influences the evolution of brain and gonad size in the male bat. (Whether this has any bearing on human evolution is hard to predict, but, ladies, you have been warned.)

Bats' porn-star credentials don't end there. It turns out they are also among a select group of mammals known to engage in oral sex. The female short-nosed fruit bat was the first to be observed performing fellatio on her partners; then a few years later the males got in on the action, when cunnilingus was documented in another fruit bat species, the aptly named Indian flying fox. The scientists were surprised. The only other mammals they could find evidence of regularly indulging in this type of sexual behavior were primates, and they reportedly "held many meetings to discuss the functions."

Their eventual conclusion was that oral sex prolonged intercourse and therefore increased the chance of fertilization. It may also, particularly in the case of the Indian flying fox, be a way for the male to suck out a competitor's sperm prior to launching his own.

The resulting academic paper (complete with graphic blow job-by-blow job descriptions of bat sex that would make a vicar blush) concluded that further voyeurism would definitely be required: "Observation at close-range is needed to find out whether the male's tongue enters the vagina or not." Well, quite.

OVER THE CENTURIES, the bat's disturbing physiology aroused almost as much bad press as its diet. Among the more ridiculous rumors was

La.rv.

Elpertílio.Arefto.Uefptíliões
bñtpedesfieutanes:z carêt cau
oa qm funtagreftes,Jpazq3 ale
funtmembranalef id eftcozio indiftincte
z fi caudam babuiffent.motum ala3 pro

Half a dozen bats earn their "ham mouse" moniker as they congregate greedily over some cured meat in the early German encyclopedia *Hortus Sanitatis* (1491), feeding the paranoia that bats were after people's bacon

the accusation that bats were after your bacon. This widespread medieval myth was described in one of the earliest encyclopedias of the natural world, *Hortus Sanitatis* ("The Garden of Health"), published in Germany in 1491. Nestled among solemn accounts of dragons and the diagnostic virtues of urine there is even a helpful woodcut illustrating half a dozen bats hovering greedily around a dangling ham. Indeed, the German name for bat is *Speckmaus,* literally "ham mouse."

In the early nineteenth century, when other researchers were wrestling with big questions like the origin of the species and the ordering of the elements, a pair of German scientists, Kuhl and Hermann, contributed to the march of scientific progress by attempting to sustain a cage of would-be bacon thieves with a daily ration of rashers. These were resolutely rejected by the incarcerated beasts, which starved to death after a week. The bats' terminal hunger strike did, however, satisfy the Germans that their beloved hams were safe.

The bats' reputation only got worse when news emerged of species that sustained themselves with something far more sinister than cured meat: other animals' blood. The association of bats with a certain Transylvanian count would prove to be the nail in the coffin for human-bat relations.

The first reports came from the sixteenth-century explorers of the New World who returned to Europe with vivid tales of bloodthirsty beasts. In 1526, Gonzalo Fernández de Oviedo y Valdés, the Spanish writer and historian who slurred our friend the sloth, described bats that "suck such a great amount of blood from the wound that it is difficult to believe unless one has observed it." The following year, the Spanish conquistador Francisco Montejo y Álvarez and his troops were said to be victims of "a great plague of bats which attacked not only the beasts of burden but the men themselves, sucking their blood when they were asleep."

These early descriptions win prizes for melodrama, if not for plausibility. For a start, vampire bats don't suck blood; they lap it up from an open wound like a cat drinking milk. And while they can consume close to their own body weight during a single thirty-minute bender, they are the size of a mouse, which means their liquid diet amounts to little more than a spoonful—an insignificant portion of the several quarts of blood swilling around a mammal the size of a human. They also rarely attack people; they are much more likely to feed on domesticated animals like cattle or chickens. The macabre tales of the conquerors of the New World put these fiendish beasts on the map, but it would be some time before the bats acquired their infamous blood-sucking handle.

The word *vampir* has its roots in Slavic language and means "blood drunkenness," but the mythology can be found in a number of ancient cultures from Babylon and the Balkans to India and China, suggesting a fear deeply embedded in the human psyche. These supernatural ghouls roamed the night draining humans dry of life and possessed the power of transformation, but they never took the form of a bat. The vampire's alter egos were an unlikely bunch that would struggle to get cast in a modern-day horror flick: horses, dogs and fleas were common, but vampires were also said to take the form of watermelons and household tools.

Nevertheless, during the late seventeenth and eighteenth centuries, vampirism became an obsession in much of Eastern Europe. This was a time of mysterious scourges like the black plague and smallpox whose deadly effects were often believed to be the work of the "living dead." Newspapers reported vampirism as fact; sovereigns sent delegates to investigate "vampire epidemics" in Hungary, Prussia, Serbia and Russia. It was only a matter of time before the reports of the mythological bloodsuckers and the authentic blood-lovers collided.

The taxonomist Carl Linnaeus officially anointed bats with their infamous title in 1758. He described *Vespertilio vampyrus* as the species that "draws blood from the sleeping at night" in the tenth edition of his classification bible, *Systema Naturæ*. There followed a flurry of similarly named bats from all over the world: *Vampyressa* (1843), *Vampyrops* (1865) and *Vampyrodes* (1889) were all variations on the theme. Johann Baptist von Spix, curator of zoology at the Bavarian Academy of Science, showed a little more imaginative flair when he named the new species of bat he'd collected in Brazil *Sanguisuga crudelissima*—"a bloodsucker most cruel." It became known colloquially as the "long-tongued vampire." Spix said he'd seen them "hover like phantoms in the profound darkness of the night."

The problem is that not one of these bats had ever tasted so much as a sip of blood. They were all innocent fruit eaters, damned to carry the misleading baggage of a sanguivorous name for scientific eternity.

It was easy enough to identify the scene of a vampire bat attack—the anti-coagulant in the bat's saliva made the wound bleed sufficiently to leave a big, bloody clue in the morning—but laying hands on the culprit during its nocturnal drinking binge was significantly harder. Taxonomists in Europe, trying to identify the true vampire from a lineup of dried specimens and highly unreliable "eyewitness" evidence of

New World explorers, made a fateful mistake: they assumed the largest bats they were sent must be the bloodsuckers (they were, in fact, benign vegetarians). When a Spanish naturalist finally got his hands on an actual vampire bat, no one believed him that it drank blood.

Cartographer and military commander Félix de Azara succeeded in collecting the creature that would become known as the common vampire bat in Paraguay in 1801. Azara was a talented amateur naturalist who discovered hundreds of new species. He had, however, dared to criticize the great Buffon, calling the Comte out for what he believed to be "vulgar, false and mistaken notices" in the aristocratic Frenchman's *Histoire Naturelle.* This impudence did not sit well with the European natural history set of which Buffon was self-appointed royalty. When Azara claimed to have laid his hands on this infamous and elusive "biter," the scientific establishment quickly dismissed him. The animal in question was named *Desmodus rotundus* after its fused front teeth. No mention was made of its gory diet.

Vampires had penetrated the popular consciousness by the early nineteenth century. The embrace of batlike wings, batlike movements and—eventually—actual bats in these Romantic Gothic horror stories made for considerably scarier monsters than blood-sucking melons. And they breathed new life into old legends. The most popular was Bram Stoker's *Dracula,* published in 1897, but all through the century bat fact and vampire fiction had become entwined, with an innocent fruit bat cast as the evil villain. This led to some disappointing cases of mistaken identity.

In July 1839, the Surrey Zoological Gardens had scored a major coup by procuring a "Vampyre" that they proudly boasted was "the first living specimen ever seen in England." The gardens were home to an eclectic menagerie once owned by the impresario Edward Cross, who had relocated his animals from the Strand in London after an elephant with a bad case of toothache had murdered its keeper. He was looking to start afresh after the scandal, and the gardens were hoping to draw a crowd with this display of the infamous bat. The press, however, were somewhat underwhelmed by the behavior of the legendary beast. "Although this bat is the Vampyre Bat, to whom so many bloodthirsty traits have been attributed," went one dispatch, "his appearance is, by no means, ferocious." The bat was "docile" and "appeared fond of being noticed." The greatest disappointment: it would "devour cherries"—and nothing else. This Vampyre was a fruit bat, after all.

Deliciously bloodthirsty "scientific accounts" of vampire bats had promised a "formidable species" that flew into bedchambers to attack their unwitting inhabitants. "If they find any part of the body exposed, they invariably fasten upon it, insinuate their aculeated tongue into a vein, with all the art of the most experienced surgeon, and continue to suck the blood till they are satiated," said one of the more popular animal encyclopedias of the period. The author then turned his sharp eye to the details of a vampiric episode: "It frequently occurs that persons, when awaked from their sleep (through loss of blood), have not sufficient strength left to bind up the orifice. The reason why the puncture is not felt is that, whilst the Vampyre is sucking, it continues to fan the air with its wings in such a manner that the refreshing breeze lulls the sufferer into a still deeper sleep." This is what the public expected to see.

Had the Surrey Zoological Gardens got their hands on a genuine vampire, it would not have subscribed to the encyclopedia's Gothic fantasy any more than the fraudulent fruit bat. But its authentic behavior is arguably far creepier.

Vampire bats tend to approach their victim not from the air but by stalking them from the ground. They use their exaggerated winged hands to drag themselves forward while bouncing along on a pair of stunted rear legs. While this may sound awkward, they can move surprisingly fast on foot. One inspired scientist, for instance, placed vampire bats on a treadmill, where he clocked them at an impressive top speed of more than 6.6 feet per second (five times faster than a sloth at full tilt). They can also launch themselves vertically into the air from the ground, like a Harrier Jump Jet, which helps make for a swift getaway.

Then there's the issue of their iconic liquid diet. The bats tend not to feed indiscriminately, but instead use special infrared sensors in their nose to detect the heat of blood that's pumping close to the skin—and therefore easier to access. Favorite hotspots for incision are fur- and feather-free; think feet (ticklish), ears (annoying) and anuses (good Lord). The bats will return to feed on the same spot several nights in a row, guided by their unique ability to listen out for and memorize a preferred victim's breathing patterns.

Bloodthirsty bum-biters that stalk their prey by listening to them breathe may sound even more evil than Dracula himself. In actuality, vampire bats are one of the animal kingdom's most magnanimous animals.

Life as a flying mammal is an energetically expensive existence, for which an exclusively hematophagous diet provides far from the ideal fuel. Blood consists of 80 percent water and absolutely no fat. Vampire bats have a specially adapted digestive system that allows them to eliminate excess water quickly by urinating as they feed, which enables the bat to consume maximum blood protein in one sitting without its stomach exploding. But with no fat, and no opportunity to build up fat reserves, they must feed at least every seventy hours or they die. And it's not as easy as you might imagine to find an exposed foot or anus (consider what hooves and tails were invented for). Up to 30 percent of bats fly home empty. Two consecutive nights of feeding failure means almost certain starvation.

One of the world's leading experts on bats, Professor Gerald Wilkinson of the University of Maryland, has discovered an adaptation that improves their odds: the bats have evolved a food-sharing system in which victorious vampires vomit up congealed blood for their hungry neighbors. This *Exorcist*-style projectile puke might turn your stomach, but it's nothing short of a lifesaver to a starving bat. Wilkinson told me that the "bats appear to be *competing* to give blood." More curiously, they regurgitate for roost mates that are not even family members, and are more likely to share with previous blood-regurgitating partners than their own relatives. "Relatedness has got nothing to do with it," he said. "A bigger predictor of sharing behavior is whether they have been helped by that individual before." In this caring, sharing, blood-spewing community, bats form strong and meaningful bonds with one another. "You can think of them almost as friends."

The bats' reciprocal altruism flies in the face of conventional models of what biologists call *kin selection*, in which animals are expected to do favors only for those with whom they share genes. Examples of reciprocal altruism are extremely rare in the animal kingdom. Wilkinson pointed out that "outside of the primates—baboons and chimps—they are very hard to find." So it's not just that pendulous penis that primates have in common with the bats. "Vampire bats are similar to primates in that they also use grooming to establish social bonds, and that influences who helps whom and the formation of coalitions," he said.

If you are a vampire bat, it'll take more than a bit of casual grooming to get an invite to the blood-spewing party, however. One of Wilkinson's PhD students recently found that the bats' bonds take a considerable time to form. Bats placed in an enclosure and forced to roost with complete strangers don't start sharing food for a good two years. "They don't just trust anyone," Wilkinson said. Vampires have an extraordinarily long lifespan for an animal of their size—thirty years compared to two to three years for a similar-sized mouse. It would seem that they use their time to develop a wide circle of longstanding, charitable friendships.

ONE OF THE REASONS that bats slipped so easily into vampire mythology is that they appear to possess occult powers, at least when viewed from a human's sensory perspective. Their spooky ability to navigate in the dark branded them familiars of witches, which meant it was genuinely scary for a single lady to have bats visiting her home during the Middle Ages. In 1332, Lady Jacaume of Bayonne, in France, was publicly burned after her neighbors spied "crowds of bats" fluttering in and out of her property.

Various bits of bats were often used in sorcery. Shakespeare's witches in *Macbeth* included "wool of bat" in their famous incantation, but they weren't following the usual recipe; bats' blood was the perennial favorite of any genuine wannabe sorceress. It was also a key ingredient in "flying ointment," which was said to help keep a witch from bumping about in the dark on her broom. Despite the popularity of this concoction from the fifteenth to the eighteenth centuries, no woman is likely to have left the ground. However, other ingredients in the salve, like belladonna, might have made certain ladies *feel* like they were flying, thanks to their psychotropic effects.

It took science rather a long time to figure out the source of the bat's seemingly supernatural skills. Not all bats echolocate; most of the fruit-eating mega-bats use their eyes to navigate. Those bats that do, form a complex sonic map of their surroundings by listening to the echo of their calls and judging distances by the quality of sound waves reflected back at them. This is mind-bending enough, but the concept was especially hard for scientists to get their heads around since bats seem to be silent. They are, in fact, screaming their heads off about twenty decibels louder than the speakers at a Black Sabbath concert (fronted by the bat-biting rocker Ozzy Osbourne). It's simply that bats' high-pitched squeals fall almost exclusively outside the human range of hearing and are therefore inaudible to us.

It wasn't until the 1930s, when a Harvard biologist named Donald Griffin collaborated with an engineer to build a special sonic detector, that we were finally able to eavesdrop on the bats' silent screams and the notion of their having some kind of supernatural "sixth sense" was finally swept away. This was a great if late development for bats, as the cryptic chiropterans had suffered well over a hundred years of torture in the efforts to extract their bio-sonar secrets.

Their trials had begun in the eighteenth century, with the Italian Catholic priest Lazzaro Spallanzani, whose résumé reads like the work of a biological sadist. He sliced off the heads of seven hundred snails to find out if they would regenerate (he claimed they could) and forced ducks to swallow hollow glass beads to understand the pulverizing action of their gizzards. He was also the first person to resuscitate the strangely indestructible microorganism known as a tardigrade, the only animal to survive freezing, radiation and the vacuum of space (and Spallanzani's insatiable curiosity). With such a keen interest in dissecting and resurrecting life, it is perhaps little wonder that he sought the sanctuary of ecclesiastical sponsorship. The Church helped finance his experimental endeavors and also provided him with some degree of absolution.

In 1793, aged sixty-four, Spallanzani turned his sights to the bat's knack for navigating in the dark. He had noticed that his pet owl became totally disoriented and crashed into walls if the candle that lit his room was blown out. Why, he wondered, did a bat not do the same? To find out, the padre sharpened up his clippers for a set of excruciating experiments.

It all started relatively innocently. Spallanzani fashioned a selection of tiny hoods for his subjects, choosing different types of cloth and styles designed to obscure the bat's vision to varying degrees. To increase the challenge, he released them in a room containing a homemade obstacle course of long twigs and silk threads that hung from the ceiling. The hoods left the bats disoriented in flight, a bit like his owl, but Spallanzani wasn't sure if this was because the bats were flying blind or because their hoods were too tight. So he took the next logical step and blinded the bats.

"We can blind a bat in two ways," he wrote in one letter of his long and gory correspondence with his Swiss collaborator, Professor Louis Jurine (pronounced with a regrettably silent "J"). He went on to inventory his medieval torture-house methods: "burning the cornea with a thin, red-hot wire, or . . . pulling the eyeball out and cutting it off."

Keep in mind, this was a man who, in an effort to understand his own digestive juices, had once taken to swallowing cloth bags of food attached to long strings, which he could hoik back up after a decent period of stewing. What were a few bats' eyes in the quest for knowledge? Especially when the results would prove so thrilling:

> With a pair of scissors I removed completely the eyeballs of a bat . . . [when] thrown into the air . . . the animal flew quickly . . . with the speed and sureness of an uninjured bat . . . my astonishment at this bat, which absolutely could see, although deprived of its eyes, is inexpressible.

The revelation was indeed nothing short of a miracle. Especially given that Spallanzani had filled the sockets of the bat's eyes with hot wax and covered them with tiny leather goggles, just for good measure.

Having deduced that the blinded bat couldn't possibly be navigating by sight, Spallanzani and Professor Jurine proceeded to creatively eliminate the animal's remaining senses, one by one.

First, they tackled the sense of touch, since blind humans at the time were rumored to find their way "unharmed through the streets of a city" by sensing "changes through their skin." Spallanzani used a pot of furniture varnish to "coat the whole body of a blinded bat, including snout and wings." Not surprisingly, the heavily lacquered animal at first struggled with flight, although it soon "regained its vigour" and flew without restraint. Leaving nothing to chance, the priest repeated the experiment with yet more lacquer. "It is to be noted," he wrote Jurine, "that a second and third coat of varnish does not hinder the normal flight of the animal."

The attempt to eliminate the bat's sense of smell led to his first big experimental setback. "I stopped up the nostrils," Spallanzani informed Jurine, "but the creature soon fell to the ground, overcome with the difficulty of breathing." The awkward problem of the bats' needing to breathe forced him to improvise. He next fastened "small fragments of sponge" soaked in strong-smelling salts in front of the bat's nostrils. They "flew as freely as ever."

The results from the test for taste were more perfunctory: "The removal of the tongue produced no result."

One thing did make a difference to the bat's flight: obliterating their capacity for hearing. This the Catholic priest achieved through a variety of means worthy of the Spanish Inquisition. He tried cutting or burning off the bat's ears, sewing them shut, filling them with hot wax and piercing them with "red hot shoe nails." The last approach finally proved too much for the bat, which "fell down in a perpendicular direction when thrown into the air." It died the following morning, raising uncomfortable questions about whether the not inconsiderable discomfort involved in these experiments was in fact the cause for the bat's blundering flight.

Spallanzani, never to be defeated, came up with yet another creative solution, crafting custom-made miniature ear trumpets out of brass.

These could either be filled with wax (eliminating sound) or left empty (providing a control).

It was the experiments with these tiny ear trumpets that finally gave him the confidence to declare that bats need to be able to hear in order to see in the dark. The only issue was the apparently silent nature of bats in flight, which troubled the priest no end. "But how, if God love me, can we explain or even conceive in the hypothesis of hearing?"

In the end he supposed that the sound of bats' wings might somehow be reflected by objects "and that they judge the distances from the quality of this sound." How was he to know that bats are actually screaming louder than a fire alarm, at a frequency outside human perception? The study of sound at this time was still very much in its infancy, although significant breakthroughs were only around the corner.

Given the ingenious and fastidious nature of the experiments, not to mention the eye-popping sacrifices made by the bats, it is unfortunate that this work was mostly ignored by the scientific establishment. But it was. For the next 120 years, it was widely asserted that bats navigated not by sound, nor even by sight, but by touch.

This conviction can be traced to one man, the esteemed French zoologist and anatomist Georges Cuvier (the more famous brother of the beaver-keeping Frédéric). For reasons best known to himself, Georges was unconvinced by the methodical mutilations of Spallanzani and Jurine. In 1800, without making a single experiment himself, the Frenchman authoritatively declared in the first of his epic, five-volume review of comparative anatomy: "To us the organs of touch seem sufficient to explain all the [obstacle avoidance] phenomena which bats exhibit."

Cuvier's star was very much on the rise, and his word final. Amid the turmoil of revolutionary Paris, the ambitious scientist had the ear of Napoleon, the self-appointed emperor of France, who'd charged him with developing a national science program. Lone voices of dissension, such as British physician Sir Anthony Carlisle—who after his own experiments concluded that bats avoided obstacles "owing to extreme acuteness of hearing"—went largely unheard. The more typical attitude was expressed by one George Montagu, who in 1809 sarcastically asked, "Since bats see with their ears, do they hear with their eyes?"

While such academic jeering must have frustrated our dedicated "bat men," generations of bats were required to undergo a further century of torture and mutilation. Investigators around the world took to replicating the dynamic duo's experiments. Countless more bats were shaved and covered in Vaseline, their eyes glued shut or gouged out, their ears excised or plugged with cementlike substances. Every attempt failed to produce a conclusive result. Ultimately, salvation (for both the bats and the thwarted researchers) arrived from a remarkable source: the sinking of the *Titanic*.

Sir Hiram Stevens Maxim was an American-born, British-naturalized engineer with a particular flair for innovation. He dreamed up gadgets for everyone—the world's first portable automatic machine gun (for the boys), hair-curling irons (for the girls), automatic fire sprinklers (for the conscientious) and self-resetting mousetraps (for the less so). His most elaborate project was a steam-powered flying machine that briefly "flew" before crashing in 1894. Disasters perhaps weighed heavily on Maxim's mind after that point. For when the *Titanic* had its catastrophic collision with an unseen iceberg in 1912, he was spurred to devise a means to prevent similar tragedies, and his inspiration came entirely from bats.

"The wreck of the *Titanic* was a severe and painful shock to us all," he wrote at the time. "I asked myself: 'Has science reached the end of its tether? Is there no possible means of avoiding such a deplorable loss of life and property?'" The inventor didn't deliberate for too long. "At the end of four hours it occurred to me that ships could be provided with what might be appropriately called a sixth sense, that would detect large objects in their immediate vicinity without the aid of a searchlight."

Maxim borrowed the idea of the sixth sense from a close reading of Spallanzani's long-dismissed work. The engineer was struck by the soundness of the proposal that bats must navigate by hearing. He decided they must be listening for the reflected echo of the sound of their wings, and that the reason they seemed to be silent was that the sound they made was outside the frequencies audible to humans. Here, Maxim made a key mistake: he assumed the bats' pitch was *below* our hearing range rather than above it. He also incorrectly assumed the source of the sounds was the bats' wings instead of their mouth and nose. But he was right that the sounds are *outside* our hearing range. This was the crucial missing piece of the puzzle and paved the way for the next wave of thinking. A few years later, the British physiologist Hamilton Hartridge suggested that bats made inaudible high-frequency sounds. Then it was only a matter of time before their secret sonar was sussed out.

Man-made sonar was worked out first, however. Shortly after Maxim went public with his proposal, the British meteorologist Lewis Richardson and German physicist Alexander Brehm filed separate patents for an acoustic navigation system that—like bats'—detected sound reflected off objects to judge their size and relative distance. In 1914, an iceberg was successfully detected from a distance of nearly two miles in a field test. If Cuvier hadn't obscured Spallanzani's gruesome research, nautical sonar might have been invented a good decade earlier, perhaps saving the fifteen hundred people who drowned with the doomed ocean liner. We will never know how the course of history might have been changed.

SIR HIRAM MAXIM wasn't the only maverick inventor to have his imagination fired up by bats, but he may have been the sanest. A harebrained scheme to blow up Japanese cities during the Second World War using thousands of bats as incendiary devices was significantly less successful.

Dr. Lytle S. Adams, a sixty-year-old dentist from Pennsylvania, was driving home from a holiday in New Mexico on December 7, 1941, when news broke that Japan had attacked the US fleet moored at Pearl Harbor. Shocked and outraged, the dentist began to ponder a plan for American retaliation. He recalled the clouds of bats leaving the famous Carlsbad Caverns he'd seen earlier in his vacation. What if tiny bombs were strapped to thousands of bats and released into a

Japanese city? The bats would naturally seek refuge in the nooks and crannies of homes, where the bombs would explode, killing unsuspecting Japanese citizens as they slumbered.

What could possibly go wrong?

Well, quite a lot. The technology of the time had yet to contrive a bomb lighter than a can of beans, which would be a struggle for an animal the size of a mouse to lift off the ground, let alone carry in flight over a great distance. Remote detonation was also very much in its infancy. Then there was the awkward problem that bats, unlike other military animal conscripts such as pigeons, porpoises and dogs, cannot be trained to follow commands.

But despite these screaming flaws, the dentist's idea was given the green light for funding by the US military. Adams, it turns out, had friends in some very high places. This dentist who dabbled in invention had persuaded First Lady Eleanor Roosevelt to check out his earlier idea to deliver and pick up mail from a plane without ever having to land, and somehow his demonstration had made a sensible impression. So when Adams detailed his incendiary-bat plan in a letter to Franklin D. Roosevelt, it didn't immediately wind up in the bin where it belonged. Instead, the letter was forwarded to the National Research Defense Committee—the group from which the Manhattan Project was spun off—with a personal note of recommendation. "This man is not a nut," the president wrote before concluding, somewhat rashly, "It sounds like a perfectly wild idea but is worth looking into."

Adams's "Proposal for Surprise Attack" had, in fact, more than a whiff of crackpot about it. He promised to "frighten, demoralize, and excite the prejudices of the people of the Japanese Empire," while at the same time providing a purpose for the planet's "despicable" winged mammals. "The lowest form of animal life is the bat, associated in history with the underworld and regions of darkness and evil. Until now reasons for its creation have remained unexplained," he wrote. "As I vision it the millions of bats that have for ages inhabited our belfries, tunnels and caverns were placed there by God to await this hour."

Adams did indulge one tiny, niggling concern in his letter to the White House. It was important to keep in mind that his "practical, inexpensive" plan to destroy "the Japanese pest" might easily "be used against us if the secret is not carefully guarded." The batty plan was duly stamped "top secret" and assigned the suitably sci-fi code name Project X-Ray. A crack team was assembled of senior army types, arsenal experts, engineers and biologists, including Donald Griffin, the Harvard scientist who had decoded the bat's echolocation riddle in the 1930s. Together, they set about vaulting the scheme's more vertiginous hurdles.

The first stage was to capture thousands of Mexican free-tail bats from caves in the southwest US, where they roosted in astonishing numbers—tens of millions. Then a bomb had to be developed that was light enough for the teeny half-ounce bats to transport. And in a quintessentially American twist, parts for the diminutive bomb were manufactured in a factory owned by celebrity crooner Bing Crosby.

Bats from this enormous network of caves had actually been re cruited in previous wars. Or rather, their feces had. Anyone who has ever visited a busy bat cave is painfully aware of the high levels of nitrogen contained in their guano; the intensely acrid smell of ammonia hits your throat as soon as you enter. When the Confederate states ran

Developing a bomb small enough to be carried by a four-ounce bat was just one of the many problems in conscripting *Chiroptera* as airborne incendiary devices during the Second World War; their inability to follow orders being another (with suitably explosive results).

short of supplies during the American Civil War, the Southerners improvised by extracting this nitrogen to manufacture their explosives.

With bats and bombs sorted, it was time to conjoin them. The miniature explosives were to be attached to the bats with simple twine, with the idea being that the bats would "fly into hiding in dwellings or other structures, gnaw through the string, and leave the bombs behind." The clever scientists assumed they could control the animals using their biology. They placed the bats in refrigerators, forcing them into hibernation, for easy handling and transportation, but timing their thaw proved tricky. Several early tests with dummy bombs were a dud because the bats woke either too late (causing them to plummet ingloriously to the ground with their cargo) or too soon (allowing them to escape the base).

Undeterred, the scientist ran a test using real incendiary devices in June 1943, less than two years after Adams had hatched his scheme. Things did not go as planned. A report on the experiment by one Captain Carr stated somewhat evasively that "testing was concluded . . . when a fire destroyed a large portion of the test material." What the captain failed to mention was that the barracks, control tower and countless other buildings at the Carlsbad auxiliary field station were spectacularly set ablaze by a bunch of escapee bat bombs. The need to maintain military secrecy prevented civilian firefighters from entering the scene. People were forced to retreat to a safe distance and simply watch as the fires leapt from building to building, destroying most of the base. As a final insult, a couple of winged missiles went AWOL, taking up roost under the general's car before exploding.

The project never really recovered from this ignominious retreat. It limped on for another year, under new guidance from the Marine Corps, but in 1944 it was finally cancelled. Bing's bombs were a dud. Having set up some thirty tests and spending a couple of million dollars, the Americans put their focus behind developing a bomb that exploited the power of atoms—which proved to be quite a bit easier to control than bats.

CHAPTER 7

FROG

Order Anura

*It is a most singular thing, but, after a life of six months' duration,
frogs melt away into slime, though no one ever sees how it is done;
after which they come to life again in the water during the spring,
just as they were before. This is effected by some occult operation
of Nature and happens regularly every year.*

—Pliny the Elder, *Naturalis Historia*, AD 77–79

I spent much of the millennium year in search of a mythical under-water monster with an enigmatic name—*Telmatobius culeus*, a.k.a. the aquatic scrotum frog.

I first heard about this baggy beast while staying with a well-connected conservationist in Uruguay. He told me that back in the 1960s, a friend of his, Ramón "Kuki" Avellaneda, had been cruising the vast expanses of Lake Titicaca, high up in the Andes on the border of Bolivia and Peru, in a mini-submarine with none other than Jacques Cousteau. They were on a fruitless search for lost Inca gold, but as consolation discovered gigantic aquatic frogs, which the conservationist assured me were the size of small cars.

Frogs are my favorite animal. Having made the evolutionary leap from water to land, they are, to me, the original explorers. They have overcome their inherently vulnerable biology to colonize some of the most uninhabitable corners of the earth through some truly nifty adaptations. The sixty-seven hundred or so known species of frog include those that secrete their own sunscreen, others that make their own antifreeze and some that can even fly. The earliest amphibians were genuine giants that dined on baby dinosaurs and measured over thirty-two feet long. Perhaps Kuki and Cousteau had discovered a relic beast—an amphibian Nessie at the bottom of the world's highest lake.

I managed to track down Kuki spending his golden years basking among the beautiful near the swish seaside resort of Búzios, in Brazil. Deaf as a post from his years underwater, he communicated on the phone via his son, who relayed the somewhat disenchanting information that his father's aquatic scrotum frogs were merely the size of dinner plates, not cars. I struggled to hide my disappointment.

Telmatobius was actually first discovered way back in 1867. Its laughable Latin name refers to its somewhat saggy scrotal appearance, which won't win any beauty contests but allows the frog to pull off the kind of endurance trick the great Houdini could only dream of.

Lake Titicaca is an unforgiving home. Over thirteen thousand feet above sea level, the sun is fierce and the air is thin. This is no place for a delicate-skinned, cold-blooded amphibian. But *Telmatobius* survives by living almost entirely underwater, protected from the harsh UV rays and dramatic fluctuations in temperature by a big wet blanket. It rarely surfaces and breathes almost exclusively through its skin, which has evolved into copious folds that drape around its scrawny frame to maximize the surface area. When it's in need of more oxygen, instead of surfacing to gulp air like a normal frog might, *Telmatobius* does push-ups on the bottom of the lake to increase the circulation of fresh oxygenated water around its droopy dermal pleats.

When Cousteau explored the lake in his mini-sub in 1969, he reported "thousands of millions" of these big amphibians, which he said typically measured some twenty inches long. Today, local fishermen report that Cousteau's giants have long since vanished, and even their diminutive descendants are becoming harder to locate.

These days, the best place to find an aquatic scrotum frog is inside a blender in downtown Lima. These wrinkly amphibians are the key ingredient for a traditional form of Peruvian backstreet Viagra,

Telmatobius culeus fortunately has no idea that he's been named after his saggy scrotal appearance and is quite likely to end his life in a blender as a form of backstreet Viagra.

popular throughout the country but especially in the capital. That's how I ended up begging a taxi driver I met at the airport to race me across town, in the two hours between my connecting flights, to take me to his favorite frog juice bar. As we screeched through the city at breakneck speed, I suddenly realized that my insistence on sampling the cabbie's preferred amphibian aphrodisiac might be interpreted as a come-on, so I tried to keep the conversation science based. This was something of a challenge given my broken Spanish, his lack of English and the frog's somewhat suggestive name.

We got to the juice bar, which was little more than a ramshackle room opening onto a bustling market street, and I caught my first glimpse of the legendary beast. This behemoth ball bag of the deep turned out to be a small, speckled, muddy-green frog with sad bulbous eyes staring out of its dirty glass tank.

The taxi driver ordered his regular Friday afternoon pick-me-up and, with the deftness of Tom Cruise in *Cocktail*, the no-nonsense woman behind the bar whipped a forlorn-looking frog out of the tank by its feet, whacked its head on the counter, peeled its skin like a banana and popped it in a Moulinex with some herbs and honey.

With a discernible twinkle in his eye, my guide handed me the resulting frog shake to taste. In the interests of journalism, I took a small sip. It tasted sweet and creamy and not at all froggy. It was actually quite pleasant, until I thought about what was in it, but drinking it failed to make me feel in the slightest bit sexy.

Although numerous amphibians secrete chemicals that are proving to be extremely useful to science, as with so many traditional remedies, *Telmatobius* is unlikely to offer any genuine medicinal value. It's a cultural thing. Among the Andean people, these frogs have long been affiliated with fertility, and frog imagery dates back to before the Inca ruled the region.

Other cultures had their own take on frogs. In medieval England, placing a frog in your mouth wasn't considered an aphrodisiac but rather an excellent form of contraceptive. It's hard to understand how this could have worked, although it would no doubt have deterred being kissed. In 1950s China, the Communist government's minister of health promoted swallowing a handful of live tadpoles as the preferred method of birth control. (This was a distinct improvement on an ancient Chinese prescription that recommended frying the tadpoles in mercury first, which would prevent pregnancy quite effectively by poisoning everyone involved.) The minister proceeded to run serious efficacy tests of the polliwog pregnancy preventers on mice, cats and humans. But 43 percent of the women tested fell pregnant within four months, and in 1958, live tadpoles were officially declared to have no contraceptive power, presumably much to the relief of women (and tadpoles) all over the country.

FROGS HAVE BEEN WORSHIPPED as fertility gods for at least five thousand years. The Aztecs had a giant toad called Tlaltecuhtli as an earth-mother goddess who embodied the endless cycle of birth, death and rebirth. Their neighbors in pre-Columbian Mesoamerica worshipped an even older amphibian deity by the name of Centeotl, the patron of childbirth and fertility, which took the rather disturbing form of a toad sporting a row of massive udders. On the other side of the world, in ancient Egypt, the goddess of fertility and birth, Heqet, was depicted as a frog.

The most probable source of these wide-ranging myths is the frog's habit of breeding explosively—a process no less dramatic than it sounds. Their survival tactic is to overwhelm predators by gathering in immense numbers to collectively produce so many eggs they couldn't possibly all be eaten. These gatherings can be quite a sight—a writhing mass of amorous amphibians clinging to each other in twos, threes and more, for days on end.

Since almost all amphibians must breed in water, these frog orgies frequently coincide with annual rains or flooding events that are also important for human farmers. The ancient Egyptians, for example, relied on the yearly flooding of the Nile to support their agriculture. As the floodwaters receded in the spring, they left behind a rich black soil that nourished crops—and thousands of horny frogs. The fecundity of the frogs, the earth and the people were thus entwined in people's minds.

But the big mystery was where all those frogs came from to begin with.

Their sudden mass appearance puzzled the ancient philosophers, who ventured that this sexy profusion had erupted out of the earth itself, the frogs somehow produced from the life-giving alchemy of water mixed with mud, much like eels. Aristotle's theory of spontaneous generation was liberally applied to a wide range of animals that, similar to frogs and eels, lacked obvious sex organs or underwent an incomprehensible metamorphosis. Such notions had been bumping around China, India, Babylon and Egypt for some time, but it took Aristotle to pull them all together into one seriously considered, but nevertheless ill-conceived theorem.

Like most of his teachings, Aristotle's theory was received with great reverence. In addition to unraveling the mystery of eels and frogs, his formulation provided an explanation for how maggots suddenly swarmed in rotting meat and the unnerving appearance of intestinal

worms in human excrement. Naturalists who followed in his foot-
steps, including Pliny the Elder, ran with the idea, adding to Aristotle's
list of animals and ascribing the spontaneous generation of insects out
of everything from "old wax" and "slime of vinegar" to "damp dust"
and "books." Specific dead big animals were believed to give rise to
specific small animals: horses transmogrified into hornets, crocodiles
into scorpions, mules into locusts and bulls into bees. Carcasses were a
popular "generator of life" in every form imaginable.

Every natural philosopher worth his salt was keen to get in on the
spontaneous generation game. The German Jesuit Athanasius Kircher
recommended a raft of recipes, some as simple as a pot noodle, to
the readers of his 1665 opus *Mundus Subterraneus*. To create frogs, for
example, you simply collected clay from a ditch where frogs had lived,
incubated it in a large vessel while adding rainwater and—presto—you
got a jar of instant amphibians.

Since some species of frog are known to hibernate in mud during
times of drought, there is a chance that someone may have once "created" a frog in this manner. But the same is unlikely to be true of
the imaginative instructions laid out by Jan Baptist van Helmont, a
seventeenth-century Flemish chemist, and the Gordon Ramsay of
the spontaneous generation set. His cordon bleu creations included a
method for generating poisonous predatory arachnids: take one brick,
make a deep hole and fill with basil, cover with a second brick and leave
in the sun. In a matter of days, "fumes from the basil, acting as a leav-
ening agent, will have transformed the vegetable matter" filling your
home with "veritable scorpions." To create mice, you placed some
wheat and water in a flask and covered it with the skirt "of an unclean
woman," and twenty-one days later you had a small rodent companion. Recipes for puppies might have proved a little more popular.

The theory of spontaneous generation was so widely accepted that
when the great British doubter of ancient myths Sir Thomas Browne
dared in 1646 to challenge whether mice could really be formed in this
way, he was ridiculed. "He doubts whether mice can be procreated of
putrefectation!" said one irate Aristotelian believer. "So he may doubt
whether in cheese or timber worms are generated; or if butterflies,

locusts, grasshoppers, shellfish, snails, eels and such are procreated of putrefied matter . . . to doubt this is to question Reason, Sense, Experience."

The arrival of the microscope in the mid–seventeenth century opened up tiny new worlds for cynics like Browne to explore. A modern band of microscopists and experimental biologists set about to dismiss the archaic, occult thinking as they conducted the first real scientific investigations. Among their leaders was the Italian naturalist Francesco Redi, who took it upon himself to challenge Aristotle's longstanding theory in what must be the stinkiest set of experiments ever undertaken (with perhaps the notable exception of Audubon's putrefying pig).

Over the course of a sweaty Italian summer in 1668, Redi got his hands on as many animal carcasses—from frogs to tigers—as he could source. He then assiduously followed the instructions for spontaneous generation laid out by the various natural philosophers, turning his home into a malodorous kitchen for the creation of life. No matter how bizarre or rank the methods, Redi treated them with solemn sincerity, testing each several times to see if they might hold the secret of genesis. For example, Redi followed the directions laid out by his fellow countryman, Giambattista della Porta, wherein "the toad is generated from a duck putrefying on a dung-heap"—not just once but three times. Sadly, the trials "brought no result." He was forced to report that Porta, "otherwise a most interesting and profound writer," had in fact "been too credulous."

Whatever fetid flesh he used, the only animals that Redi could generate were maggots and flies. "I continued similar experiments with the raw and cooked flesh of the ox, the deer, the buffalo, the lion, the tiger, the dog, the lamb, the kid, the rabbit; and sometimes with the flesh of ducks, geese, hens, swallows, etc.," he said. "Finally I experimented with different kinds of fish, such as sword-fish, tuna, eel, sole, etc. In every case, one or other of the above-mentioned kinds of flies were hatched."

This led Redi to take his own leap of imagination, a leap obvious to us but radical back then: that the flies hanging about the meat may in fact be the progenitors of the maggots. "Having considered these things, I began to believe that all worms found in meat were derived directly from the droppings of flies, and not from the putrefaction of the meat," he wrote. "I was still more confirmed in this belief by having observed that, before the meat grew wormy, flies had hovered over it, of the same kind as those that later bred in it."

At this point Redi began one final experiment to test his suspicions. It must have been an absolute stinker:

> I put a snake, some fish, some eels of the Arno, and a slice of milk-fed veal in four large, wide-mouthed flasks; having well closed and sealed them, I then filled the same number of flasks in the same way, only leaving these open. It was not long before the meat and the fish, in these second vessels, became wormy and flies were seen entering and leaving at will; but in the closed flasks I did not see a worm, though many days had passed since the dead flesh had been put in them.

The simple genius of this experiment—which proved that those carcasses that were protected from resting flies failed to produce maggots, while those that were left exposed were soon swarmed—was the beginning of the end of the theory of spontaneous generation.

Alas, out of its ashes a new but equally erroneous dogma emerged. Enter the "preformists," who had the goal of explaining the generation of *all* animal life. They believed that every living being developed from a miniature version of itself, called the *homunculus*, which was contained within the animal's seed, and that germination merely reflected an increase in the mini-me's dimensions. The preformists fell into two opposing camps: the "ovists," who thought that the homunculus was contained within the female's egg, and the "spermists," who believed it resided within the male's semen.

The idea that both sperm and egg were necessary for life was very much in the minority. That finally changed when, in the 1780s, our favorite scissors-wielding biologist, Lazzaro Spallanzani, proved otherwise.

Spallanzani believed the frog's amorous couplings would betray the secrets of all conception. With frogs, fertilization happens externally, so the act of conception could be more easily observed and, most crucially, controlled.

But back then, even this basic truth was contentious. Carl Linnaeus was on the record saying, "In Nature, in no case, in any living body, does fecundation or impregnation of the egg take place outside the body of the mother." So Spallanzani got out his scissors and started interfering with frog sex to check the Swede's claims. He grabbed solitary females, sliced them open and extracted their unborn eggs. These, he observed, never developed into tadpoles, growing instead into "an offensive putrid mass." Eggs released while the female was embraced by the male, in contrast, always developed into tadpoles. This was proof that conception must be happening externally and, although the male appeared to be doing little more than holding on tight (frog semen is invisible in water), Spallanzani suspected he must be contributing *something* to the process. He just needed to find it.

To do so, he borrowed an idea from a French scientist, René Antoine Ferchault de Réaumur, who thirty years previously had gone to great lengths to hunt for whatever substance, if any, a male frog emitted during copulation. This he did in the most ingenious fashion—by forcing the frogs to wear homemade amphibian underpants, which served as a form of full-body prophylactic. Fortunately for Spallanzani (and for us), this fastidious French scientist kept meticulous minutes of the various prototypes for these pants.

"On March 21st we put a pair of pants made of bladder," Réaumur wrote in his notes, "very tight pants, that sealed off the hind end." The animal bladder was nice and stretchy and easily slipped on the amphibians. But once the frogs got in the water it became "too soft and

One for the spermists—a homunculus nestled inside a spermatozoon, as imagined by Nicolaas Hartsoeker in his "Essay de dioptrique" (1694).

floppy" and started to disintegrate. He wasn't sure if "the frog was adequately covered," so these original undies were scrapped.

Waxed taffeta, a waterproof material used to make umbrellas, proved a more durable choice. Unfortunately, it lacked the necessary elasticity to ensure a snug fit. With palpable frustration the Frenchman noted: "Having made the pants and put them on, the frogs abandoned them in front of me." The leg holes he'd made were too big, and much to his dismay, the frogs could remove the pants by pushing their legs up through the holes and jumping clean out of them.

Réaumur was nothing if not resourceful. He resolved the issue by bespoke-tailoring the pants with tiny braces that slipped over the frog's shoulders and held the getup snugly in place. Reading through

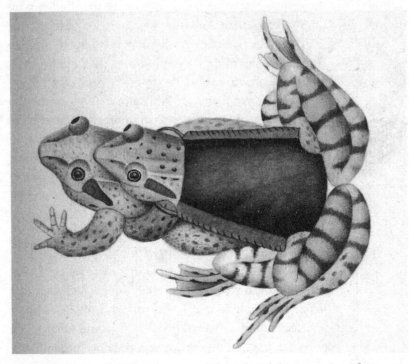

The French scientist Réaumur was so pleased with his invention of custom-made amphibian underpants that he commissioned an artist, Hélène Dumoustier, to preserve this froggy fashion statement in perpetuity, braces and all. And who can blame him.

Réaumur's notes, Spallanzani was inspired to fashion similar attire for his own amorous amphibians. "The idea of breeches, however whimsical and ridiculous it may appear, did not displease me, and I resolved to put it into practice. The males notwithstanding this encumbrance, seek the females with equal eagerness, and perform, as well as they can, the act of generation."

After the frogs had done the deed, Spallanzani carefully removed their pants and peered inside to inspect his catch. Unlike his French forebear, the Italian priest succeeded in collecting a few precious blobs of semen, which he promptly wiped on a batch of unfertilized eggs. These eggs went on to develop into tadpoles, suggesting that the residue from the frog pants was indeed essential for fertilization.

The systematic Spallanzani went on to confirm that nothing else could have brought the eggs to life, wiping a series of spawn with blood, vinegar, spirits, wine (various vintages), urine and the juices of lemon and lime. He even tried bringing them to life with electricity. All were fruitless, in reproductive terms.

SPALLANZANI'S SARTORIAL ADVENTURES with frogs were a key step in solving the mystery of fertilization. Less than a hundred years later, amphibians were back in the lab, once again being used to divine whether fertilization had taken place—this time not in a frog, but in a human.

It may sound suspiciously like bogus medieval folk medicine, but from the 1940s through to the 1960s the world's first reliable pregnancy test was a small, bug-eyed frog. When injected with a pregnant woman's urine, the amphibian didn't turn blue or display stripes, but it did squirt out eggs eight to twelve hours later to confirm a positive result.

There was no such thing as a home testing frog. The injecting was done by professional pregnancy testers who whiled away the hours in the basements and adjoining buildings of many a hospital and family-planning clinic alongside tanks of these prognostic frogs. I discussed the process with Audrey Peattie, an effervescent eighty-two-year-old former frog tester from Hertfordshire, who told me about her three-year turn with the slippery hoppers at Watford hospital.

Working in a lab full of urine and amphibians was an unusual occupation for a young woman in the 1950s. Whereas most of Audrey's
female friends left school to become secretaries, she headed off to
Watford aged seventeen for a career that was a little more "peculiar"
and "embarrassing to explain," but which she relished nevertheless.

"We did about forty tests a day. The frogs were rather slippery but
you got hold of them between their legs and injected them under the
skin and into their fleshy thighs," Audrey recalled. "You then popped
them into a numbered jar, left them in a warm place overnight and
examined them in the morning to see if they'd laid eggs. If the frog
had laid only a few eggs, or no eggs at all, we repeated the test with
another toad, just to be sure it wasn't wrong. But the toads were practically never at fault."

According to Audrey, these frogs were "not the kind that creep
around your garden" but the altogether more exotic African clawed
frog (*Xenopus laevis*), an ancient species of aquatic amphibian from
sub-Saharan Africa. Armed with long talons, flat bodies and what appears to be Frankenstein stitches down each side of their body, they're
not exactly pretty. Their bulbous eyes also lack eyelids, which means
they bob around in the water with a somewhat ominous stare that
follows you around the lab.

The frog's knack for detecting pregnancy was discovered by the
British endocrinologist Lancelot Hogben while he was stationed at
the University of Cape Town in the late 1920s. Hogben had previously
used European frogs in his studies of hormones, but in South Africa he
began experimenting with the local fauna. He found that *Xenopus*, just
like today's chemical pregnancy tests, displayed a dramatic response
to the presence of human chorionic gonadotrophin, or hCG, the hormone released after a human egg is fertilized. Hogben recognized the
toad's potential as a pregnancy test to be a "godsend." He was so en-
amored with the amphibian that he later named his house after it.

The "Hogben test," as it became known, quickly replaced the far
less reliable "rabbit test," which involved injecting a rabbit with urine
and then dissecting it several hours later to examine its ovaries for any
sign of eggs. "Imagine keeping enough rabbits to do forty tests a day!"

Audrey Peattie (right) at work in the 1950s at the family-planning lab in Watford hospital, where she was engaged in the slippery business of encouraging toads to tell her whether women were pregnant or not.

Peattie said. The frogs had the distinct advantage that they could be reused.

Another plus of the Hogben test was that the critters involved were small and could be kept in tanks while awaiting their rendezvous with an uncertain woman's hormonal sample. After doing a predictive turn, each frog would get "about three weeks off." During this time "they just swam about" and "got fed chopped liver." Then they would get called up for duty.

The African clawed frog single-handedly revolutionized pregnancy testing, destigmatizing a process once associated with animal death and making the service practical on a larger scale than ever before. But this doesn't capture the whole of their scientific importance. Hundreds of thousands of frogs were exported from Africa to testing labs in Europe and America, where their presence attracted the attention of other scientists, particularly those in the emerging field of developmental biology—the intellectual descendants of Spallanzani—whose attempts to map the growth of embryos required copious quantities of eggs for experimentation. The amphibians they had been using bred seasonally, which severely restricted their work. Here was a frog that could be induced to lay tens of thousands of eggs on command, simply by injecting it with hCG (which is kind of like a very modern recipe for spontaneous generation). As an added bonus, Xenopus eggs were unusually large—ten times the size of a human egg—making them ideal for microsurgery and genetic manipulation. The tadpoles were also conveniently transparent, allowing developmental biologists to observe the internal mechanics of metamorphosis into adults. To cap it all, the adults were highly resistant to disease and could live in captivity for up to twenty years. It was a marriage made in scientific heaven.

Xenopus would go on to join the ranks of the mouse and the fruit fly as one of the most intensively studied model organisms on the planet, with live colonies in labs in forty-eight countries spread across five continents. By the 1980s, Xenopus had become the world's most widely distributed amphibian. It had been decoded, dissected and documented inside and out. It was the first vertebrate to be cloned and had even been to space.

THERE WAS ONE CRUCIAL THING that scientists still didn't know about it—something they didn't discover until, unfortunately, it was too late. It turns out this globetrotting frog doesn't travel solo.

In the late 1980s, herpetologists started noticing something very odd. Populations of amphibians in Australia and Central America were disappearing, often quite suddenly, from pristine environments and leaving behind no dead bodies. It was as if they had simply vanished into thin air. Amphibians had been around for sixty-five million years, and had survived the meteor that exterminated the dinosaurs, a series of ice ages and quite a few dramatic climate fluctuations. What could be killing them off in such massive numbers now?

After years of intense speculation, the culprit was eventually identified: a primitive waterborne fungus, *Batrachochytrium dendrobatidis*, a.k.a. Bd or amphibian chytrid fungus. Chytrid fungus infects the frogs' skin—an especially sensitive organ that they use to breathe—causing a disease that prevents the absorption of oxygen and essential electrolytes. It eventually results in cardiac arrest.

Over the next thirty years, scientists watched in horror as the chytrid fungus cropped up on every continent on Earth—except Antarctica, where there were no amphibians. Its spread triggered a catastrophic decline in or complete annihilation of at least two hundred species. Even in these extinction-heavy times, this amphibian apocalypse has been described as the most spectacular loss of vertebrate biodiversity in recorded history due to disease.

Where had this frog-killing fungus come from, and how had it spread so far *and* so fast?

A few years ago I traveled to Chile, a country hit hard by chytrid, to meet with Dr. Claudio Soto-Azat, an up-and-coming scientist determined to find the answer to these big questions and save his country's amphibians. Claudio is one of those people who brims with joy, a great asset to have when dealing with a subject as depressing as the mass extinction of your favorite animal.

Compared with its neighbors, Chile doesn't have a dizzying catalogue of amphibians (a mere fifty species), but the ones that it does have are almost all unique to that country. That's because Chile is essentially a long, thin island isolated by desert to the north, glaciers to the south, ocean to the west and the Andes to the east. So despite being part of a vast frog-filled continent, Chile's amphibians have evolved in a bubble, which makes them especially vulnerable to extinction.

I joined Claudio on an expedition in search of one of the country's most fabulous freaks, the incredibly rare southern Darwin's frog, which was discovered by the big beard himself in 1834 on his epic five-year *Beagle* voyage. What makes this frog so extraordinary is that it has eschewed conventional pond-based metamorphosis for something more sci-fi: after mating the male guards the fertilized eggs until they are close to hatching, then gobbles them up. Six weeks later, like a scene out of *Alien*, he barfs up baby frogs. He is the only male animal other than the seahorse to give birth, albeit through his mouth.

Claudio and I flew to a tinpot airport in Patagonia, little more than a dusty airstrip scratched into the earth and framed by snow-capped peaks. He pointed out the border with Argentina—a wobbly metal gate on a dirt track surrounded by fields and mountains. From this isolated spot, we drove four hours to reach the forests that Darwin's frog calls home—a surreal mixture of dense bamboo, giant rhubarb plants with leaves big enough to live under, wild fuchsia bushes popping with hot-pink flowers and lofty trees dripping long, wafting tendrils of pale green moss. A thick mist hung in the air. It was all very *Lord of the Rings*.

The good news was that we quickly found our frog, something of a miracle given that it's only one inch long and camouflaged to look like a bamboo leaf, complete with a long, thin nose that mimics the stem. The bad news was that the swabs Claudio took from our little green friends came back as positive for the chytrid fungus when tested in the lab.

Contracting chytrid is not an automatic death sentence for a frog. It is a mercurial killer with infuriatingly unpredictable efficacy. Certain amphibians, for whatever reason, appear to have immunity and are able to resist its suffocating grip. We could only hope the southern Darwin's frog, with its largely terrestrial life, would be safe from life-threatening levels of the fungus's waterborne spores.

A cousin, the northern Darwin's frog, had not been so lucky. Despite a less remote home range, closer to the capital of Santiago, these equally bizarre, mouth-brooding frogs have not been seen or heard from for thirty years. Claudio presumed them to be extinct in the wild and fingered the fungus. And he had a pretty good idea of how it arrived in Chile.

The next stop on our amphibian apocalypse tour was a small farm in Talagante, a semi-rural area twenty-five miles north of Santiago. Claudio wanted to investigate reports of an alien invader, his suspect number one in the spread of the frog killer.

We arrived in the heat of the day and were greeted by a farmer called Jurgen, an elderly man with watery-blue eyes, a long white beard and a warm smile. He handed us a bag of tadpoles and a bucket of toads and told us his property had been infested with these strange frogs since the late 1970s. With obvious emotion in his voice he recalled how, two years after the toads arrived, he experienced his first "silent spring," completely absent of the chirrupy singsong of his beloved native amphibians. He went searching for them in places that were previously inky black with tadpoles and found nothing. They had completely disappeared.

I peered into the bucket and was greeted by a familiar blank, bug-eyed stare. Well, hello, *Xenopus*, what the hell are you doing here?

Claudio explained that this alien's invasion was thought to be an unlikely addition to the pantheon of crimes committed by Chile's famous dictator, General Augusto Pinochet. The story goes that shortly after the military junta took over the Santiago airport in 1973, a consignment of *Xenopus* had come in by plane, destined for lab work in the capital city. Unfamiliar with the protocol for greeting a bunch of foreign frogs, the soldiers released them. They and their descendants have been on the run ever since.

The same characteristics that make *Xenopus* an ideal lab animal also make it the textbook invasive species. It is a highly adaptable, disease-resistant and prolific breeder. Females can reproduce all year round, generating up to eight thousand eggs annually. Today, the head count of African clawed frogs in Chile is "millions if not billions. It's impossible to say but it's a huge number. In one small lagoon, for example, they estimate the population is twenty-one thousand individuals."

The toads have been found as far as 250 miles from Santiago. They seem to be advancing from the capital at a speed of roughly six miles per year. During periods of heavy rain, they make mass migrations, pushing deeper into new territory. In one such period, Claudio told me, a ranger observed a biblical scene of two thousand toads crossing a road.

African clawed frogs are voracious predators. They swallow everything in their path, laying waste to populations of indigenous fish, frogs and tadpoles. This unstoppable amphibian army boasts a secret weapon with which to annihilate the indigenous froggy fauna: many of the fugitive *Xenopus* in Chile test positive for the chytrid fungus to which they appear to have developed immunity. But the extent to which *Xenopus* have contributed to the chytrid pandemic has not been clear until quite recently.

In a cunning piece of scientific detective work, Claudio, along with a handful of other international investigators, tested a pickled lineup of *Xenopus* specimens from museums around the world. They discovered individuals collected as early as 1933 had been carrying the fungus—the earliest recorded incidence of the disease, locating it to the time when the frogs were first exported out of Africa to be used as pregnancy testers. Many of these frogs didn't stay locked up in a lab. When the Hogben test was usurped by the little blue strip, thousands of redundant toads were released by well-meaning staff, who hoped to give them their freedom after a lifetime of good service. Countless others have escaped from labs or been liberated as unwanted pets. Invasive populations of African clawed toads have been recorded advancing across four continents, and the latest research has linked some of these invasions, such as those in Chile and California, with the arrival of chytrid fungus and the disappearance of native frogs. Other widespread invasive amphibians like the American bullfrog—which is farmed internationally for its fleshy legs—may have also carried the disease, but it appears as though the toads' exodus from Africa could have been the trigger for the global outbreak.

We owe much of our understanding of fertilization and embryonic development to *Xenopus*, but in gaining this knowledge, we have inadvertently caused the disappearance of fabulously freaky species like the mouth-brooding northern Darwin's frog. The high-altitude home of the aquatic scrotum frog has been infected by the fungus too. "We are living in a time of homogenization of our wildlife," Claudio said.

"Thanks to globalization and human population growth there are even more chances for the translocation of wildlife around the world—and their diseases."

The modern industrial farming method of creating dams favors the lifestyle of *Xenopus*, which thrives in still and stagnant water. The farmer we visited, Jurgen, had one small irrigation pond on his farm that he referred to as "the hell hole."

In five thousand years the frog had come a long way. The ancient Egyptian farmers may have worshipped it for its fecundity, but in Jurgen's eyes, *Xenopus* was more akin to an embodiment of the curse bellowed by the Omnipotence in Exodus: "I will plague your whole country with frogs. The Nile will teem with frogs. They will come up into your palace and your bedroom and onto your bed, into the houses of your officials and on your people, and into your ovens and kneading troughs."

Looking at Jurgen's putrid pool of fetid water squirming with frogs, I thought how their sudden appearance might be more plausibly explained by Aristotle's outlandish spontaneous generation than the bizarre globetrotting, pregnancy-testing, pathogen-infested, explosive-breeding truth.

CHAPTER 8

STORK

Species *Ciconia ciconia*

In diverse sorts of fowl their absence is such, that we know not
whither they go, nor whence they come, but are as it were
miraculously dropped down from Heaven upon us.
—Charles Morton, "An Essay into the Probable Solution
of This Question: Whence Comes the Stork," 1703

On an otherwise unremarkable morning in May 1822, Count Christian Ludwig von Bothmer was out hunting on the grounds of his castle in Klütz, Germany, when he shot a rather unusual white stork. The bird in question had already suffered a near fatal attack—not from a gun, the count's weapon of choice, but a wooden spear, a yard in length, which skewered the stork's long, thin neck like a kebab. The count brought the spiked stork to a local professor, who deduced that the primitive weapon, roughly hewn from exotic hardwood with a simple iron blade at the tip, had been thrown from "the hands of an African." How extraordinary. The animal had not only survived being lanced, but then had somehow summoned the strength to fly thousands of miles to Europe wearing its extravagant neck piercing, only to be shot dead by the count upon arrival.

What was undoubtedly a rather bad day for the bird turned out to be a significantly better day for science. The examination of the

This famous *Pfeilstorch*, or "arrow-stork," shot down in Germany in 1822, provided irrefutable evidence that birds migrate over Africa—and wears a suitably weary look for its heroic sacrifice to science.

impaled stork helped to solve one of the longest-running mysteries of the natural world: the seasonal disappearance of birds.

THE WHITE STORK, *Ciconia ciconia*, is a conspicuous creature. Adults have arresting white and black plumage and stand well over three feet tall on long scarlet legs. Everything about them is showy, not least of all their gigantic nests that teeter in plain view on top of the tallest buildings in towns all across Europe. They also have a habit of clattering their big scarlet bills together in clangorous glee when greeting their partners during their nuptial dance at the start of each spring.

These big boisterous birds became firmly embedded in ancient folklore. They were harbingers of spring, talismans of good luck. All across Europe, people encouraged storks to nest on their roofs in the belief that the birds brought harmony, health and prosperity to a household.

How lucky the human hosts felt once storks had taken roost is debatable. Generations of white storks will return to the same nest, adding to it every year. Nests are mostly made of twigs, but an inventory of objects found in storks' nests included "a ladies glove, a man's mitten, horse droppings, an umbrella handle, a child's skittle ball and a potato."

And the resulting nests are big. Stork supernests can weigh up to two tons and measure over eight feet deep, a structure that modern, let alone medieval, residential engineering would struggle to withstand. Yet nests survived for centuries (albeit with more than a little human help). One perched on top of a tower in the German town of Langensalza stood proud for four hundred years. A document from 1593 noted the sum given for the repair and upkeep of the nest, suggesting that the human residents were under a certain obligation to maintain the two-ton pile of twigs weighing down their home.

STORKS ARE MOST FAMOUSLY ASSOCIATED with fertility. In many European countries, it was believed that a couple with storks nesting on their house would shortly be blessed with a baby. Even today in Germany, a wooden model of a stork carrying a bundle in its beak is often placed outside a house where a baby has just been born, and pregnant women are referred to as having been "bitten in the leg by a stork." The prevalence of this belief can cause some confusion. The US TV network Fox News recently reported on a German couple who attended a fertility clinic because they had failed to have children, only to be told that in order to have a baby, they had to actually have sex first. They thought the stork was enough.

The big white birds' reputation as baby bringers has its roots in pagan culture. The birds reappear every spring, and spring was generally a bumper time for births. The summer solstice on June 21 was a traditional pagan holiday of marriage and fertility. Many romantic liaisons would take place then, and nine months later, coinciding with the appearance of the storks, their babies would also arrive. The two events became connected, with the result that people thought the storks brought the babies.

These birds really know how to make their presence felt—which makes their absence, come autumn, all the more evident. After a summer spent raising their chicks in plain view, they vanish for several months, reappearing early in the new year. It seems obvious to us today that the birds have migrated over twelve thousand miles to southern Africa in search of more favorable feeding opportunities. But the nature of the stork's disappearing act, along with that of all other migratory birds, was the subject of several of natural history's most enduring prevarications.

Aristotle was the first person to ponder seriously why some birds apparently vanished at the arrival of a new season only to reappear, as if by magic, at the start of another. The great thinker hedged his bets with a trio of theories. Some birds, he supposed, such as cranes, quails and turtle doves, sought out warmer climes during the cool European winter. He even noted that the birds fattened themselves up prior to their journey. He should have stopped with migrations as his one winning idea. But perhaps because of the fantastic nature of this biological endurance act, the grandfather of zoology felt the need to come up with two more explanations, which were not only wrong, but lingered in the scientific realm well beyond their sell-by date.

Aristotle's more creative alternative was transmutation. His epic *Historia Animalium* declared that certain species of bird transformed into different species from one season to another. Summertime garden warblers, for instance, became wintertime blackcaps, and winter robins became summer redstarts. These birds shared a superficial likeness, similar in size and coloration, and, most suspiciously to the philosopher, they were never completely present at the same time (much like Clark Kent and Superman). Redstarts migrate to sub-Saharan Africa at a time when robins, who breed farther north, come to winter in Greece. So Aristotle concluded they were avian shape-shifters.

The philosopher's idea of transmutation was tame in comparison to the fantasies that followed in its wake. Alexander of Myndus, another Greek, asserted four hundred years later that elderly storks transform into humans. This was reported with solemnity by Claudius Aelianius as fact. "In my view, this is not a fairy tale," he stated in his second-century animal encyclopedia, *De Natura Animalium*. "If it were, why would Alexander tell it to us?" he wrote. It's worth noting that in the same book, Aelianius—who must surely rank as one of history's most gullible encyclopedists—also described sheep that changed color according to which river they drank from, tortoises that absolutely "hated" partridges and octopuses that grew as big as whales.

Storks weren't the only birds thought to transmutate. There were even more fantastic stories involving the barnacle goose, a bird that we now know migrates from obscure Arctic seas to the coasts of Britain every winter. Its most northerly breeding grounds on the high cliffs of Greenland were never observed by the European authors of the medieval bestiaries, who instead promoted the rather unlikely story that barnacle geese grew out of the rotting timbers of ships.

"Nature produces them against nature in the most extraordinary way," wrote the twelfth-century chronicler Gerald of Wales. "They are produced from fir timber tossed along the sea." The medieval clergyman claimed to have actually witnessed the bird's phenomenal genesis during an expedition to Ireland. "Afterwards they hang down by

CHAP. 171.

Of the Goose tree, Barnacle tree, or the tree bearing Geese.

Britannica Conchæ anatiferæ.
The breed of Barnacles.

Barnacle geese were said to grow from trees as well as from rotting timber, which meant that birds often featured in plant books and (more important) conveniently qualified for consumption in the many meat-free fast days of the medieval calendar.

their beaks as if they were a seaweed attached to a timber, and are surrounded by shells in order to grow more freely. Having thus in process
of time been clothed with a strong coat of feathers, they fall into the
water or fly freely away into the air."

What Gerald had observed were actually goose neck barnacles,
from the taxonomic order Pedunculata. As their name suggests, these
finger-sized filter feeders bear a resemblance to a small beak sitting
atop a long, naked neck as they strain out of whatever they've attached
themselves to in the tidal zone. So great was the human mind's power
of association that the esteemed sixteenth-century botanist John Gerard claimed to have cut specimens open to have inside "found living
things that were very naked, in shape like a bird." Some of these supposed beings were said to be "covered with soft down, the shell half
open, and the bird ready to fall out."

There was, however, a more convenient motivation for the popularity of the myth: it sanctioned the consumption of roast goose, when
the eating of meat was strictly forbidden. Growing out of rotting ships
meant these geese were not classed as flesh "because they are not born
of flesh," explained Gerald. Such sneaky logic allowed "bishops and
men of religion make no scruple of eating these birds on fasting days."
Given that three days of the medieval week, not to mention the whole
of Lent, were designated fast days, it's easy to see how famished men
of the cloth would be keen to propagate the fable that this big, juicy
bird was in fact a viable vegetarian menu item.

But what of the noisy white stork? Where did it go in winter?

Aristotle's third theory about the disappearing birds was less fantastic but significantly more enduring. In the *Historia Animalium* he
proposed that storks, along with a number of other species, escaped
the cold by going "into hiding," which makes them sound like some
kind of avian fugitive. Aristotle went on to explain they were in a
state of "torpor." Many mammals—which are warm-blooded, like
birds—had been observed by early natural philosophers to hibernate,
including bats, which were then often classed as birds. So why not
storks too?

It's a good question, one that modern science doesn't have a definitive answer for. It's probably down to a combination of factors: a relatively high metabolism and heart rate, and the struggle to store sufficient fat, would probably make hibernation physically challenging for the stork. Not to mention the fact that a stork lacks the equipment to dig a suitable burrow to sleep in. Instead, it possesses a pair of perfectly good wings that can carry it away to somewhere more favorable.

There are a handful of birds—hummingbirds, mousebirds and swifts—that have been discovered to enter brief spells of torpor, but so far modern science has confirmed only one species of truly hibernating bird: the common poorwill—a type of nightjar that is most common in the desertlike habitats of western North America. Some poorwill do migrate to avoid winter food shortages but find themselves competing with numerous other migrating birds in the all too popular wintering hotspot of Mexico. The rest avoid the competition by lowering their metabolism and sleeping out the winter among the rocks, an evolutionary adaptation that earned them a Native Hopi name meaning "the sleeping one."

Despite the lack of evidence of hibernation in all other birds, ornithologists argued about this from ancient times well into the nineteenth century. The bird at the center of the sleepy academic storm wasn't the stork but another well-known harbinger of spring, the barn swallow.

Aristotle claimed these tiny birds hibernated in holes while "quite denuded of their feathers." The idea that swallows would spend their winter months sleeping naked in order to survive the cold is quite silly, but the theories that followed over the course of the next two thousand years were sillier still. The Age of Enlightenment, which began at the end of the Thirty Years' War in the mid–seventeenth century and came to a close with the French Revolution near the end of the eighteenth century, saw some of its greatest minds, no less than the fathers of modern zoology, sincerely believing that swallows spent the winter hibernating at the bottom of lakes and rivers, like fish. "It appears certain that swallows become torpid during winter, and even that they pass the season at the bottom of the water in the marshes," wrote

Georges Cuvier in his highly influential nineteenth-century tome, *Le Règne Animal*.

The scissors-wielding biologist Lazzaro Spallanzani was also intrigued by the bird's Houdini act and brought his trademark sadistic flair to numerous experiments looking to find proof of it. He tried to encourage swallows to hibernate by imprisoning them in wicker cages that he buried under snow, leaving just a small hole through which the birds might breathe. The swallows skipped torpor and went straight to death in less than two days. Over in France, the Comte de Buffon attempted a similar experiment, placing swallows in an icehouse, with equally fatal results.

In America, the barbarity peaked with a physician by the name of Charles Caldwell, who, with the assistance of his "invaluable friend" Dr. Cooper, attached ballast to a pair of swallows' legs before chucking them in a river. In his manifestly sinister write-up of his experiment, Caldwell reported that the birds, which he referred to as "our two little prisoners," sank like stones, "showing the anxiety and convulsions of animals in a drowning state." After three hours, the birds were removed and the men of science attempted to resuscitate them. The swallows were in fact "reduced, not to a state of torpidity, or suspended animation, but of absolute death."

For many years, one German university offered a reward of the bird's weight in silver for each swallow found underwater and revived. The prize was never claimed. Nonetheless the legend, unlike all those experimental swallows, failed to die. How had this mad submarine swallow rumor started?

An obscure sixteenth-century Swedish bishop by the name of Olaus Magnus appears to be the culprit. Magnus did not entertain the idea of migration, despite intimations from other quarters that it might be the proper explanation. "Although the writers of many natural things have recorded that the swallows change their stations, going, when the winter cometh, into hotter countries," he wrote in *Historia de Gentibus Septentrionalibus* ("History of the Northern Peoples"), "yet, in the northern waters, fishermen oftentimes by chance draw up in their nets an abundance of swallows, hanging together like a conglomerated

The Swedish bishop Olaus Magnus was not one to let the truth get in the way of a good story. His 1555 bestseller *History of the Northern Peoples* is chock-full of ridiculous fables presented as fact, like these fishermen hauling hibernating swallows out of a riverbed.

mass." This giant bird ball, the holy Swede tells us, formed when swallows descended like tiny line dancers into the depths. "In the beginning of Autumn, they assemble together among the reeds; where, allowing themselves to sink into the water, they join bill to bill, wing to wing and foot to foot."

Unfortunately for the swallows, Magnus's magnum opus was a massive bestseller. All twenty-two volumes of his encyclopedic extravaganza mixed fact with fable, painting his frozen homeland as an otherworldly place where mice rained from the sky and giant serpents prowled the seas offshore. The Swedish bishop's sensationalist tales appealed to the new wave of readers who now had access to books following the development of the printing press in the mid–fifteenth century. His tome was translated into a dozen or so languages, spreading his fantastic legends all over Europe.

But it was the nascent Royal Society of London that gave the story of the submarine swallows a stamp of credibility. In 1666—a short six years after its founding—this fellowship of the world's most eminent natural philosophers decided to investigate the matter and establish "what truth there is . . . concerning swallows being found in winter under waters

congealed, and reviving, if they be fished [out] and held to the fire."
Their conclusion: "It was most certain, that swallows sink themselves
towards autumn into lakes." A somewhat surprising result from this
storied scientific body until you learn that the man tasked with exam-
ining the matter was not a naturalist but an astronomer whose analysis
involved little more than consulting an acquaintance of his who just
happened to be a professor at the University of Uppsala, former hometown of one Olaus Magnus. The chance of getting an unbiased answer
from an academic from Magnus's alma mater was about as likely as
finding a swallow wintering underwater. The legend was so deeply entrenched in local folklore in Sweden that even Carl Linnaeus—another
Uppsala alum—still referred to it as fact a hundred years later.

Not everyone bought the hibernation story. One fierce opponent
was Charles Morton, the Oxford-educated author of a highly respected seventeenth-century compendium of physics that was standard issue at Harvard for almost half a century. Morton pointed out
with the unflinching logic of a physicist that the freezing temperatures
and distinct lack of air meant the idea of swallows "lying in clay lumps
in the bottoms of rivers" was utterly ridiculous. He proposed the more
rational hypothesis that swallows, as well as other seasonal birds like
the stork, migrated to the moon.

"The stork, when it hath bred, and the young fully fledged . . . all
rise together, and fly in one great flock . . . first near the earth, but after
higher . . . till at last this great cloud . . . appears less and less by distance, till it utterly disappears," Morton mused, before turning to the
crux of the matter: "Now, Whither should these creatures go unless it
were to the moon?"

Whither, indeed. Morton's evidence for this celestial sojourn was a
little on the slim side. He reasoned that since no one knew the where-
abouts of migrant birds during the winter months, they simply must
be hiding somewhere off the earth. Further evidence could be gleaned
from the bird's demeanor. At their departure the bird's "cheerfulness
seems to intimate, that they have some noble design in hand, namely,
to get above the atmosphere, hie and fly away to the other world."

His outlandish hypothesis was a reflection of the times. The scientists of the seventeenth century were fascinated by the moon. Using one of the first telescopes, the Italian scientist Galileo Galilei had observed that the lunar surface wasn't smooth like a marble but was instead punctuated by mountains and valleys, just like earth. John Wilkins, a former colleague from Morton's university days and a founder of the Royal Society, wrote *The Discovery of a World in the Moone*, an enthusiastic account of how lunar geography was much like earth's, complete with seas, streams, mountains—and possibly life forms. To Morton, the moon, far from being a lifeless lump of rock devoid of atmosphere, was an alluring winter destination.

The notion of space migration had wider support, with proponents arguing among themselves via letters to the Royal Society about which celestial body was the most likely destination for the birds. The Puritan minister Cotton Mather thought the moon a little too distant. He suggested that birds instead flew "to some undiscovered satellite accompanying the earth at a near distance." Mather was famous in New England for erudite sermons, and for fanning the flames of hysteria around the Salem witch trials, including a spirited defense of "spectral evidence." So it should come as little surprise that he also believed in space-traveling birds.

Charles Morton was singular in his pursuit of the idea, however. He made painstaking calculations, which are at times quite well reasoned, on behalf of his avian astronauts, exploring the parameters of their extraterrestrial migration.

He divided the year into three parts. The round trip to the moon occupied four months, or sixty days, each way; this gave the birds four months living on earth and four months living on the moon. As the moon took one month to revolve about the earth, the birds launching straight at the moon, by Morton's reckoning, "will find it in the same line of direction, where it was when they began their journey"—which was wonderfully convenient.

Fuel for their two-month mission was provided by excess fat, and the impetus for leaving earth were changes in temperature and the availability of food. All of which holds some truth for earthbound migrating birds.

Morton estimated the distance to the moon as 179,712 miles: by no means a sorry speculation, since the closest stage of the moon's elliptical orbit is in fact around 226,000. During their lunar mission, Morton assumed the birds to be unaffected by gravity, so that they encountered no air resistance and were able to cruise at speeds of around 125 miles per hour, a fair clip compared to their more typical 20 miles per hour. Here, he failed to account for the fact that the velocity required for his space-traveling storks to leave earth's gravitational pull would be some two hundred times his calculated top speed, a feat unachievable without the help of NASA's finest rocket boosters bound to the birds' backs. Then there were those small problems of the life-sucking vacuum, intense radiation and hostile temperatures the birds would encounter in space—lethal to any animal other than the bizarre microscopic creature revived by Lazzaro Spallanzani, the decidedly indestructible tardigrade.

For Morton's generation, of course, space travel was still very much a dream. It would be three hundred years before anyone could confirm or deny the presence of storks, swallows or any other birds orbiting the earth like satellites en route to the moon. Yet, this was also the great Age of Discovery, and keen-eyed European explorers were beginning to report sightings of their seasonal homeland birds while navigating foreign seas and probing faraway lands. In 1686, for example, the survivors of a Dutch ship wrecked off the coast of South Africa reported that storks were "found during the time they are not in Holland, although not in great numbers."

Such eyewitness accounts were easily dismissed by nonbelievers. The Right Honourable Daines Barrington, a staunch anti-migration fellow of the Royal Society, was one of the most vehement voices. He relished rebuffing the explorers' claims, legitimizing his own with his imperious delivery. In response to a sighting by Sir Charles Wager, First Lord of the Admiralty, of a great flock of swallows settling on the rigging of his ship, Barrington slyly twisted the evidence to his side. It simply showed the birds were not fit for long-distance migration, he said. "They are, in fact, always so fatigued, that, when they meet a ship at sea, they forget all apprehensions and deliver themselves to sailors."

Barrington, a former judge, could argue black was white and had a frankly absurd answer for every sensible suggestion he heard. He dismissed the idea of migration on the grounds that it was too dangerous to believe, and said there were not enough witnesses of these "highly improbable" events to give it any credence. Those who suggested the birds were flying too high to be seen (which they do in order to hitch a ride on favorable air currents) were shot down for being "destitute of proof." Those who hypothesized, correctly, that migration happened at night (as a way of avoiding predators) were dismissed as ridiculous since, according to the judge, everybody knew that birds, like humans, sleep at night (also untrue).

With such rabid anti-migration bulldogs as Barrington primed to criticize all claims, what was needed to end the debate was some cold hard evidence. And that's where Count Christian Ludwig von Bothmer's elaborately pierced stork comes in. For here was a bird caught carrying an irrefutable souvenir of its stay overseas, evidence that ultimately led to a paradigm shift in ornithological understanding.

The count's heroic stork proved to be no lone freak. It was just one of twenty-five similarly stoic so-called *Pfeilstorchs*—storks skewered by arrows—to be shot down in Europe during the nineteenth and twentieth centuries. Ornithologists took inspiration from these lanced birds and started their own tagging system, using something a little more user friendly than an arrow—a branded aluminum band that could be fixed around a bird's leg. This small ring revolutionized bird studies, eventually revealing conclusive proof of the storks' and other birds' seasonal migrations.

The most significant of the early bird ringers was Johannes Thienemann, an unusually flamboyant Protestant minister from Germany. Thienemann wasn't the first person to ring birds—that prize had been won by a Danish schoolteacher a few years before—but he was the first to do so on a grand scale, and to tag birds that made long-distance migrations into Africa.

Thienemann was no bog-standard boffin. He was a colorful character with boundless enthusiasm and a knack for self-promotion, qualities that enabled him to found a brand new form of scientific establishment. On January 1, 1901, Thienemann opened the doors to the world's first permanent bird observatory. Dedicated to the study of migrating birds, the center was located in Rossitten, a remote corner of East Prussia. Thienemann's favorite study species was the white stork—the "pre-destined experimental bird"—on account of its being highly conspicuous, a predictable migrant and extremely popular with the general public.

Thienemann's infectious passion for bird watching and his PR smarts helped him to corral an army of civic volunteers across Germany to help him tag two thousand storks with a unique number and the location of their tagging. That was the easy part; the rest was somewhat out of his control. All Thienemann could do was watch the storks fly away and wait and hope that someone in the vast Dark Continent would find the bird, notice the ring and send news of this oblique discovery in such a way that it found its way back to his tagging HQ in Prussia.

The gung-ho ornithologist's grand vision was not without its opposition. Kurt Floericke, the editor of the influential science journal *Kosmos*, mounted an especially vociferous attack. The editor posited that the aluminum rings would harm the birds and described the exercise as "vain scientific humbug," predicting it would end in the "mass murdering of storks." But Thienemann wanted all the publicity he could get, even the negative sort, so that word would spread far and wide of his ambitious experiment. There were few telephones and no such thing as television; Thienemann had to rely upon international newspaper dispatches and the bureaucratic corps of the African colonies if he was to hear anything about his precious stork tags. Having some missionary or colonial officer looking out for piles of stork corpses wasn't going to hurt.

Lo and behold, the first information about a tag made its way back to Thienemann just a few months later from Africa. Along with the bird to which it was attached. Now rather dead. This wasn't quite how the minister had envisaged his tags would turn up, but it was still a kind of success.

The northern journey of the tags to Thienemann was no less momentous than their initial southern migration. They passed through the hands of missionaries, colonial officials, traders and newspaper editors before making it home to Rossitten, creating something of their own mythology along the way. Many were discovered by African hunters who thought the mysterious metal objects were of "heavenly origin." One chief was said to have carried a stork tag on the shaft of his spear as a lucky charm; he considered it sufficiently precious that it only made its way back to Thienemann after the chief's death.

Between 1908 and 1913, Thienemann received news of forty-eight recoveries, which he plotted on a map, revealing for the first time the extent of the stork's impressive migration down the Nile and onwards to the southern tip of Africa. But just as the riddle of the stork was being solved, the birds themselves began disappearing from towns and villages across western and northern Europe. And this time they seemed to be disappearing permanently.

The migrant lifestyle was becoming dangerous. Their annual flight path took them over nations engaged in the First World War, places where people were starving and eager to hunt a big bird for their pot (hence the reported increase in storks arriving in Europe with souvenir arrows). In 1930, Thienemann wrote with dismay about the decline of "our dear storks," which were suffering the effects of hunting by natives.

He also speculated that the poisoning of grasshoppers by the South African government was to blame for the dwindling numbers of his favorite study subject. He was right. Modern industrialized agricultural methods were the enemy of the white stork. It was a sorry fate for a bird nicknamed the "farmer's friend" after its gluttonous appreciation of common pests.

A solitary stork has been clocked devouring up to thirty crickets a minute; another dispatched with forty-four mice, two hamsters and a frog in a single hour. A flock is estimated to be able to hoover up a plague of Tanzanian army worms—over two billion grubs—in a day.

So the introduction of pesticides not only made the birds redundant, it also gave them a terminal case of indigestion.

Pesticides, pollution and the drainage of wetlands for farmland forced Europe's stork population into sharp decline over the twentieth century. The last breeding pair was seen in Belgium in 1895, in Switzerland in 1950 and in Sweden in 1955. Many villages were quite bereft when the storks stopped returning.

The last thirty years have seen concerted conservation efforts to bring back Europe's storks. On a torrential day in June 2016, I made an exceedingly damp trip to Diss in Norfolk to investigate one of these projects. Conservationist Ben Potterton picked me up from the railway station and drove me along the flooded country lanes to Shorelands Wildlife Gardens, home to his very own, highly imaginative scheme to reintroduce storks to the British landscape.

Ben has a magic touch with animals. He has coaxed the rarest, most reluctant breeds into reproducing and thus regularly supplies stock for zoos and conservation programs. He has a weakness for "the little brown jobs"—his term for uncharismatic species that get overlooked by the big conservation campaigns. His wildlife center is a chaotic, cacophonous place awash with obscure cocoa-colored oddities from pygmy marmosets to red-breasted geese, many of which are allowed to roam free. My first encounter was with a particularly vocal duck that was aiming to deny its nature and escape the rain. She sat with us in the shelter of the cafe as Ben outlined his vision for the future of British white storks.

Back in 2014, Ben was contacted by a Polish animal rescue center that had a surfeit of white storks needing a new home. Most were the victims of power-line electrocutions, an occupational hazard of nesting on the top of utility poles. These lame birds could no longer migrate and would perish in the cold of a Polish winter, but the rescue center had been struggling to find someone willing to take on twenty-two handicapped storks. It occurred to Ben that the Polish birds could produce perfectly able-winged babies to repopulate Britain's storkless shores. So he went to Poland, packed the birds in flat-pack wardrobe

boxes—"perfect for a stork since they can be transported individu-
ally"—and drove the birds to England.

When Potterton and I met his immigrant storks, the majority were
busy stalking the perimeter of a nearby field in a big, bedraggled group.
Two of the birds, however, had hived themselves off from the crowd
to build a nest. With great pride Ben showed me the messy pile of
twigs perched atop a muddy mound around the back of the whooper
swan enclosure. The pair's unusually terrestrial setup was home to the
first British-born chicks for some six hundred years.

Storks were last recorded nesting in Britain way back in 1416, when
they were spotted on top of St Giles' Cathedral in Edinburgh. Ben
told me their disappearance in this country was not just down to the
dangers of migration, but risks much closer to home. For, unlike other
countries in Europe, where storks were so deeply revered that people
harming them could even be sentenced to death, Britain actively per-
secuted its storks.

"There was a deliberate plan by the church and our rulers, the pol-
iticians of the time, against the stork," Ben told me. "The church did
not like that they were baby-bringers because God brought babies."
The birds were dangerously associated with un-Christian, heathen beliefs that the local Church was keen to stamp out. Unlike in the rest
of Europe, where a nest on the roof was considered good luck, in England it became a sign that someone in the household was committing
adultery. Given that the medieval punishment for extramarital sex was
at best banishment (for the man) and at worst the removal of nose and
ears (for the woman), a pair of noisy nesting storks would have been
far from welcome guests.

The storks were also under fire for their political "preferences":
there was a persistent rumor that the birds bred only in republics or
countries with no king. Religious differences played a role too. Storks
were an important symbol in Islamic culture, and it was believed the
birds migrated to Mecca just as the devout did. The Scottish writer
and traveler Charles MacFarlane, visiting the Ottoman Empire in
1823, reported that "these sagacious birds are very well aware of this

predilection" and showed their allegiance with the Muslim Turks by building their nests on mosques and minarets "but never on a christian roof!"

In England vagrant storks flying over from the continent were viewed with suspicion and shot on sight. One such insurgent turned up on the doorstep of the Norfolk home of the great myth buster Sir Thomas Browne in 1668, shortly after the English Civil War and the UK's own brief flirtation with being a sovereignless republican country. Browne took in the wounded stork, nursing it back to health with hand-fed frogs and snails, and formed a bond with the bird.

His neighbors were more wary. Hopefully the bird would not foretoken a new commonwealth, they joked nervously. The ever logical Browne dismissed these "vulgar errors" as no more than "a petty conceit to advance the opinion of popular policies," adding for good measure the list of monarchies, from ancient Egypt to modern France, where the bird was known to nest.

Ben hoped his birds would be better accepted in Norfolk today. The week I visited his wildlife center, Britain voted to leave the European Union amid hysterical fears over human immigration, and it was all too easy to imagine these immigrant birds once again being viewed as intruders. Yet it struck me that in many ways these broken Polish birds were helping to recreate Ye Olde England that the Brexiters seemed to hanker for; there is archaeological evidence that storks populated England as far back as the middle Pleistocene (350,000 to 130,000 years ago). But Ben told me once the UK leaves the European Union, rewinding efforts like his that depend on animals from overseas will be mired in bureaucracy and much harder to achieve.

The big question is whether Ben's birds will migrate to Africa. Eminent zoologists including Charles Darwin believed migration was strictly an innate instinct, but scientists now believe that for soaring species like the white stork, social learning also plays a significant role; fledglings may have an imprinted urge to fly south, but it's not terribly efficient. The intricacies of their route, most crucially where to stop off and feed, are learned by following their parents. This is a luxury unavailable to Ben's storks, but he remains optimistic. Every year the coast of Norfolk receives a few transient white storks, like Thomas Browne's, from Denmark. Ben's hope is that these Danish storks would take off on their natural migration and his birds would follow them all the way to Africa.

However, new research has revealed that Europe's storks are changing their migratory habits. Many are shunning their traditional long-haul lifestyle in favor of a more sedentary existence, staying home and eating junk food.

A couple of years ago, Dr. Andrea Flack, a researcher at the Max Planck Institute for ornithology in Germany, spent a month migrating alongside a muster of fledgling storks. "We tagged sixty juveniles before they left their nests," she told me. "I followed a flock of twenty-seven individuals, chasing them every day in my car."

With the help of local firefighters, Flack began placing small GPS locators on fledglings near her research center. This was no easy task; the possessive parents were less than keen on Flack interfering with their chicks and would attack her with their beaks, while she wobbled on the end of a very long ladder. But once the birds fledged, Flack could follow them wherever they flew.

Storks, like vultures, glide on thermal currents. So they travel during the day when the sun is high and can cover great distances under the right conditions, but not necessarily along convenient highway routes. "It was a lot of driving," Flack told me. "I'd wait until eight or nine o'clock in the evening, then I would jump in my car and drive hundreds of kilometres to catch up with them." She was often driving in the dark and into the unknown. "I'd drive along dirt roads for hours and hours and then suddenly I'd end up at a pig farm and find hundreds of storks."

Flack usually slept in her car until dawn and then tried to explain to the farmer, with whom she didn't share a language, that she wasn't *trespassing*, merely following a flock of storks who'd arrived at his farm in search of food.

As Flack traveled farther south, the human population dwindled. The birds became harder to locate. She found them in secret oases, unmarked on her map. "Especially in Spain, it's a very arid area—very dry and dusty. You drive for a very long time through nothing and then you end up at this really beautiful pond that's so green and nice and surrounded by flamingos and storks," she said. "They can find these really small ponds in the middle of nowhere." Flack suspected the birds locate these secluded spots by spotting other storks circling high in the sky, riding thermals.

Flack's unorthodox road trip revealed that white storks are truly opportunistic feeders, finding pit stops to refuel on everything from frogs to pig feed. The big white birds' gastronomic tour of Europe was full of surprises, but none more so than their final feeding joint: the giant landfills of Spain.

Spain's biggest cities—Madrid, Barcelona and Seville—are surrounded by massive garbage dumps, whose mountains of organic waste and associated insects and rodents make for the ultimate smorgasbord. "Once they reach these places, the storks stay for a couple of weeks, maybe even a month," Flack discovered, "and sometimes they don't even migrate any further."

Content with this high-calorie fast food, half of Flack's birds didn't bother pushing on to Africa. They just stayed put, eating junk food all winter. When spring came, some of her storks didn't even head back to northern Europe. They stopped migrating altogether. Given that migration is probably a series of evolutionary steps from stagnation, whereby the ancestors of Flack's storks gradually shifted either their breeding grounds or winter feeding spots to leave competitors behind or chase ephemeral food bounties, it may be that some of Europe's storks are now reversing their long-distance trajectory.

I've witnessed this profound change in the storks' behavior myself. But unlike Flack, I haven't been racing to and fro across Europe to do so. I've done it all from the comfort of my sofa in London—and so can you. The Max Planck Institute has developed the highly addictive Animal Tracker app, which receives GPS data from Flack's tagged storks, along with data for a host of other animals, and plots their movements on your phone. Since 2015, I've followed a stork called Odysseus who was tagged in Germany and appears to have turned his back on his wandering nature and name. Since arriving at the landfills of southern Spain in September 2015, he's barely moved. Occasionally he has hopped over the Strait of Gibraltar to feast at the landfills of northern Morocco, but that's it. His brother Felix, tagged in the same nest, has been doing much the same.

These new tracking technologies are ushering in a golden age of migration studies, with potential to reveal the full life journey of not just the white stork but all eighteen hundred known species of long-haul migrating bird. Some of these studies are starting to confirm otherwise inconceivable acts of migrant endurance. Swifts have recently been tracked spending ten months on the wing, eating and power-napping in the air to and from South Africa, their feet never touching the ground even after they arrive. Arctic terns win the long-distance record for an annual round trip of over sixty-two thousand miles from England to Antarctica—a journey more than twice the circumference of the planet from a bird that weighs less than an iPhone. During its lifetime this diminutive aeronaut racks up around three million flying miles, which is the equivalent of four round trips to the moon—which makes Charles Morton's space challenge seem slightly less ambitious.

Other studies, like Flack's, are exposing radical real-time changes in migratory behavior. In the UK, Aristotle's blackcaps could no longer be mistaken for transmuting garden warblers; the country's recent warmer winters, and a steady supply of food from bird feeders, have removed the urge for this traditional summer migrant to ever leave Britain's shores. Swallows are also becoming increasingly reluctant to leave the UK and return to Africa. Their numbers, along with those of dozens of long-haul bird migrants across Europe, Asia and America, are in perilous decline, thanks to the combined effects of global warming, habitat destruction, hunting and pesticides. Some scientists have suggested that long-haul migration could soon become a thing of the past. These amazing avian vanishing acts, which puzzled us for so many generations, could themselves magically disappear, just as they've finally been demystified.

CHAPTER 9

HIPPOPOTAMUS
Species *Hippopotamus amphibius*

Some say that hee is five cubites high, and hath ox-hoofs,
three teeth sticking out each side of his mouth, greater out
then any other beasts, eared, tayled, and neighing like the horse,
in the rest like the Elephant; he hath a mane, a snout turning up,
in his inwards not unlike an horse, or asse, without hair.
—Edward Topsell, *The History of Four-Footed Beasts*, 1607

In his bestselling bestiary, the seventeenth-century clergyman Edward Topsell had little trouble accepting the existence of unicorns and satyrs, which he described with great glee in his bestiary. He was, however, decidedly skeptical about the hippopotamus. And who can blame him? At the time, few if any naturalists would have clapped eyes on this African beast and the descriptions doing the rounds were decidedly fantastic.

Since the days of the Roman Empire, the hippo, *Hippopotamus amphibius*, had been portrayed as a monstrous, maned "river horse" that could "vomit fire" and oozed blood. The second-century Greek writer Achilles Tatius described it "opening its nostrils wide and snorting out a reddish smoke as if on fire." One could be forgiven for thinking he had been smoking something himself, but it's likely he'd just been perusing his Bible. The mighty, chaos-causing Behemoth from the Book of Job bears an uncanny resemblance to these ancient hippos and is widely considered to be its inspiration: "Under the lotus plants he lies, in the shelter of the reeds and in the marsh. What strength it has in its loins, what power in the muscles of its belly!" bellowed God in the telling of Job's sorry story.

The biblical description of this mythical monster no doubt added to the fanciful speculation surrounding the hippo's size and fire-breathing abilities. Accounts of blood-sweating were a more credible mistake, however, and probably based on actual observation. Pliny the Elder provided an imaginative analysis in *Naturalis Historia*, his great encyclopedia completed in AD 77:

> When the animal has become too bulky by continued over-feeding, it goes down to the banks of the river, and examines the reeds which have been newly cut; as soon as it has found a stump that is very sharp, it presses its body against it, and so wounds one of the veins in the thigh; and, by the flow of blood thus produced, the body, which would otherwise have fallen into a morbid state, is relieved; after which, it covers up the wound with mud.

What reads as a tragic tale of a self-harming hippo with weight issues was in fact a portrayal of the ancient art of bloodletting—a procedure used in the treatment of an array of ailments for almost three thousand years. If you were a feverish Greek or a bubonic medieval Brit, then your doctor's top-line treatment would be to puncture a vein and drain some of your blood. If you were *very* lucky, leeches might be used instead of a sharp wooden stick. Bloodletting was practiced by the Egyptians but, according to Pliny, that other famous resident of the Nile, the hippo, had showed the way. "The hippopotamus was the first inventor of the practice of letting blood," he claimed, not once but twice, in his encyclopedia.

We may scoff at Pliny, but the truth is that this big amphibious beast has pioneered a pharmaceutical fashion, one that's practiced in our time (and, crucially, one that actually works).

The liquid the ancients observed seeping from the hippo's hide does look remarkably like blood; it completely fooled me the first time I

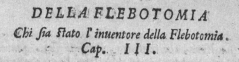

DIcono f̃naturali`, che l'inuentore della Flebotomia è ſtato l'Hippopotamo animale,che habita preſſo il fiume Nilo,di grandezza ſimile à qual ſi voglia cauallo di Friſia, & è di terreſtre,& acquatica natura,il quale,quando ſi ſente aggrauato dalla copia del ſangue; và in vn canneto,ò coſa ſimile & per iſtinto di natura ſi feriſce la vena,& ne laſſa vſcir tanto ſangue`,ſin che ſi ſenta ſgrauato : poi troua la belletta, ò fango,& iui ſ'imbelletta,& ſi ſtagna, e ſerra la ſerita della vena.

This Italian medical manual (1642) credits the hippopotamus as the inventor of phlebotomy. The animal (said to be the "size of a Friesian horse") instinctively pierces a vein, lets out blood "until it feels revived" and then wallows in the mud until the wound is healed. How terribly hygienic.

saw it. But it isn't blood—nothing like it. Instead, this crimson goop is produced by special glands tucked underneath the animal's thick skin. For many years it was thought to act like a sort of sticky red sweat to keep the hippo cool. Scientists have recently discovered it does something much more remarkable.

The slime's bloodlike appearance is the product of red and orange pigments, unstable polymers that start out clear but change shape and color as they absorb and reflect UV light. This is rather handy, because it means the hippo is essentially secreting its very own sunblock—an evolutionary adaptation for a massive, hairless mammal regularly exposed to the blazing sub-Saharan sun.

The slime is also believed to contain antibacterial agents—the reason why a hippo's war wounds hardly ever get infected despite its predilection for wallowing in water awash with its own feces. And despite their fondness for a poo party, flies tend to leave the hippo alone, suggesting this supergoop could be an insect repellent to boot.

This three-in-one formula is significantly more sophisticated than your average overpriced sun-slop from Walgreens. In fact, it's such a revolutionary substance that Christopher Viney, a biomimicry scientist in California, has been trying to turn hippo sweat into the next big thing in sunscreen. "It is the unusual combination of properties that makes it so enticing: sunblock, bug repellent, and antiseptic all rolled into one," he told me.

"Nature's most successful materials have had plenty of time to become optimized for purpose. If Nature makes a good skincare product, we will be hard-pressed to improve on it," said Viney.

There are some issues to be worked out. "The challenge," he noted, "is to get a sample that is not contaminated by feces."

Undeterred, I decided to put the professor's research to the test by smearing my own skin with fresh hippo slime. The hippo in question was an overly tame orphaned baby called Emma, resident of a rescue center in South Africa. I was feeding her when I noticed rivers of red running down her back and collecting in the folds of fat on her neck. So I decided to help myself.

The liquid had the tacky consistency of egg whites and lathered up into a creamy foam; on application, my skin quickly absorbed it. Sadly my hands were so sun worn it was hard to deduce its SPF powers, but one hand was now noticeably silkier than the other. The owner of the sanctuary was also a fan of its moisturizing qualities; she told me she regularly uses hippo goo as a lip salve and swears by it.

Emma the orphaned hippo being bribed with food, just before I swiped some of the crimson slime oozing out of her hide to test out its SPF value on my own skin.

NO OTHER MAMMAL is known to secrete its own sunblock. They don't need to; hair or fur generally does a pretty good job of protecting the skin, but this so-called river horse is so sensitive to sunlight that it evolved a sophisticated supersweat to protect it. Perhaps the answer to this unique adaptation lies hidden in the hippo's family tree.

As a zoology student in the early 1990s, I was taught the hippo was more closely related to the pig than the horse. This seemed plausible—but it was, unfortunately, wrong. Their next of kin on the taxonomic tree is a group of animals so unlikely, my former university tutor, Richard Dawkins, wrote in his book *The Ancestor's Tale* that it was "so shocking that I am still reluctant to believe it, but it looks as though I am going to have to."

The hippo's closest relatives are, in fact, whales.

For centuries, scientists had been attempting to classify the hippo by doing the usual studies of teeth and bones. But it turned out that there was another way to get a hippo to spill its shocking secret: *talking* to it.

Dr. Bill Barklow has devoted twenty years of his life to figuring out what hippos are saying to each other, making him the world's leading expert—indeed the world's *only* expert—in the lonely field of hippo communication. I met Bill in Uganda while researching animal communication for a TV series. I managed to cajole him out of retirement to teach me a few phrases in hippo. Bill does the best impersonation of a hippo's comical grunts, snorts and bellows you are ever likely to hear, perfected after many years alone with these amphibious beasts.

His twinkly blue eyes lit up when he talked about his part in the big hippo classification coup. "As a scientist you dream of that eureka moment, where you think of something no one else has ever thought of before," he told me as we motored upriver, towards the source of the Nile, on a sultry African afternoon. "But few people ever get it."

It had happened by accident back in 1987. Bill was studying loons—the North American feathered variety, famous for their haunting humanlike calls—when he decided to treat himself to his fantasy holiday: an African safari. One morning, while watching hippos for the first time, he noticed something rather puzzling. When one of the males made his raucous territorial call, within a few minutes other hippos surfaced from the depths and called back. "I thought, how can they do that? They're breaking the laws of physics!"

The density of water compared to air makes sounds above or below water reflect off the boundary, so sounds made above water can't be heard underwater, and vice versa. And yet it appeared that the hippos underwater could hear the noise the male made above, because they responded to his call. "When I got home I went straight to the library to check the literature on hippo sounds and communication. I dug deep, but found nothing."

Like any true scientist with a fresh bug in his brain, Bill waved goodbye to his loons, moved to Africa and dedicated the rest of his life to finding out how hippos defied physics. It took him the best part of ten years, but he eventually stumbled upon the answer: hippos are amphibious communicators.

When wallowing in the shallows (with just their nose, eyes and ears exposed), hippos' bellows are broadcast above water by their nostrils, but the sound is also transmitted below the water, through a large blob of blubber on their throat. Fat has roughly the same density as water, so the sound is transmitted directly from the vocal chords, through the fat and out into the river with little distortion. Submerged hippos pick up these sonic rumblings through their jawbone, which is linked to their inner ear.

Bill demonstrated this to me by playing a recording of a hippo's hello out of the large loudspeaker on our tiny boat. We had positioned ourselves as close as we dared to a small pod of semisubmerged hippos chilling out in the shallows of the river. It was late afternoon, and the air was hot and soporific. The harsh recording of the grunt roaring out from our boat shattered the peace. It was probably less than a minute before the first hippo bellowed back. Then, as if by magic, a series of solitary hippo heads bobbed above the surface of the water and joined the chorus. The effect rippled all the way downriver as far as the eye could see, with a dozen or so hippos rising out of the muddy waters, one by one, to say hello.

As Bill continued his research, he learned that hippos, known for raucous bellows as loud as thunder, carry on *most* of their communication below the surface of the water. Using a separate underwater speaker and a microphone crudely strapped to a long pole so that it could be submerged, Bill showed me how he entered the hippo's secret sonic world. He had recorded a surprising cacophony of croaks, clicks and squeaks produced by the hippos forcing air over their vocal chords or fluttering their nostrils. These decidedly unhippolike sounds, along with the way they were transmitted through fat and received through the jawbone, struck Bill as remarkably similar to a faraway group of underwater mammals: cetaceans, a.k.a. the whales and dolphins.

Bill's breakthrough caught the attention of a biologist by the name of John Gatesy who had been trying to piece together a molecular case for hippos sharing their closest common ancestor with whales for some time. This radical rethinking of the taxonomic tree had set off some suitably big waves in zoological circles. Gatesy was on the hunt for physiological evidence to support his theory.

"I had discovered that hippos make clicks or click trains underwater very similar to those that all cetaceans use for sonar," Bill said. "Hippos are also hairless, give birth and nurse underwater, and have a repertoire of underwater sounds. All of these could be synapomorphies—traits shared by both animals but no other species—indicative of a common ancestor."

The resulting "whippo theory"—which placed cetaceans and hippos in a shared evolutionary group—was met with derision by certain palaeontologists. The taxonomic tussle persisted for decades, pitting scientists studying the past against those searching for clues in the present.

One of the major issues was the incomplete fossil record of the hippopotamus line, which seemed to disappear some twenty million years ago. Fossilized evidence of ancient whales dated back much further. But in 2015, a handful of ancient Hippopotamidae teeth were found in a Kenyan canyon, providing a key stepping stone in the animal's ancestry: the molars unambiguously united the hippo with the whale family tree.

It was an ironic footnote to the hippo's heredity tale that the infamous Behemoth would turn out to be the brother of that other great chaos monster from the book of Job, Leviathan (widely considered to have been a whale). And so too that the Bible's most Brobdingnagian beasts would both have evolved from what was probably a rather diminutive ancestor, little more than the size of a spaniel.

THE HIPPO'S STRONG LINK with myth and legend led to such confusion over the centuries that the celebrated French naturalist Georges-Louis Leclerc, the Comte de Buffon, attempted to wipe the slate clean with his description of the hippo when he sat down to write his magnum opus, *Histoire Naturelle*, published in 1795. "Although this animal has been celebrated from the earliest ages it was imperfectly known to the ancients," stated Buffon. "It is only towards the sixteenth century that we have been able to procure any accurate information on the subject."

The natural history illustrators of the past struggled with the hippo as much as they did with the sloth, as is evident from this bizarre hairy-lipped and human-eyed beast—the "heppepotame" that featured in *The Gentlemen's Magazine* (1772).

Buffon's haughty tone is a little unfortunate, given that almost everything he went on to say about the hippo was wrong. It's worth looking at his claims in some detail, however, because they influenced thinking about the animal for such a long time.

To begin, he observed that "the hippo swims well and eats fish." Wrong. Hippos are vegetarians, and they can't do so much as a doggy-paddle. Instead, rendered almost weightless by water, they bounce along the riverbeds, doing a sort of submarine moonwalk.

He went on: "His teeth are very strong and of so hard a substance that they strike fire with a piece of iron. This is probably what gave rise to the fable of the ancients that the hippo vomited fire." Nice try, but also wrong. Hippo teeth have been prized as ivory since Egyptian times and, although harder than an elephant's tusks, they are still easily carved by hand and have the advantage of not turning yellow with age. In fact, if the Comte had lost his own teeth, he might have found himself wearing a pair of hippo dentures; they were all the rage in the 1700s—even George Washington had a set. (Though as far as I know, no one ever reported the first American president "vomiting fire" through them.)

There was more. "Thus powerfully armed he might render himself formidable to every animal," Buffon wrote, "but he is naturally gentle." Again, wrong, very wrong. Hippos are famously bad-tempered, highly territorial brutes with absolutely no problem using their massive tusks for battle. They weigh much the same as a family car and, despite their size, can accelerate almost as fast, easily outrunning a human. Their charging capacity combined with a penchant for attacking boats has led to a fearsome reputation as Africa's most dangerous animal.

Finally, Buffon put forward that "these animals are confined to the rivers of Africa." Wrong again, although not even Nostradamus could have predicted the latest twist in the hippo's inscrutable evolutionary tale: that a remote corner of Colombia would become a haven for hippos in the twenty-first century.

A COUPLE OF YEARS AGO, I flew to Medellín to investigate. Nestled almost five thousand feet up in the Andes, Colombia's second city is surprisingly cool and damp for a spot so close to the equator. I was met by Carlos Valderrama, a handsome thirtysomething veterinarian who has come to know hippos rather intimately.

Our safari began with a four-hour drive through lush emerald hills, down into the Magdalena River Valley and deep into cowboy country. The trek gave Carlos plenty of time to fill me in on the backstory.

Back in 2007, the Colombian Ministry for the Environment began receiving calls from rural Antioquia reporting sightings of a very peculiar creature. "They said it was really big, with small ears and had a very large mouth," Carlos said.

The locals were scared and so Carlos, a specialist in animal-human conflict, was sent in. He found himself faced with the task of explaining to bemused locals that this strange beast was an animal from Africa.

An equally confused Carlos wondered where the hippos had come from. The locals all told him the same thing: Hacienda Nápoles.

Situated halfway between Medellín and Colombia's capital, Bogotá, and sprawling over nearly seventy-five hundred acres, Hacienda Nápoles was built by the infamous drug baron Pablo Escobar. It was from here that he controlled the export of 90 percent of America's cocaine, a trade that made him one of the richest men in the world, estimated at one time by *Forbes* magazine to be worth more than $3 billion. The ranch was his personal playground. Here he threw lavish parties, amassed his collection of vintage cars and built a dinosaur park for his son, complete with life sized asbestos dinosaurs. Like many other megalomaniacs, Escobar also fancied having his very own menagerie.

Legend has it that Escobar got his hands on a massive Russian cargo plane that he ordered flown to Africa and, like some modern-day, twisted Noah, filled with (sedated) illegal wildlife. The "Ark" then had to fly back to Colombia before the animals woke up, a task made significantly trickier when he discovered the runway at the hacienda, built for small private jets loaded with cocaine, wasn't long enough. He ordered that it be rapidly extended to ensure the plane and its drowsy passengers could land.

Over the years, Escobar smuggled in lions, tigers, kangaroos and four hippos—three females and one very randy male he called El Viejo, a popular Colombian Mafia nickname meaning "Old Man." The hippos were introduced into a small lake next to the hacienda's mansion. And they're still there. Although when I paid a visit to meet El Viejo, his harem was considerably bigger than three.

Escobar was shot by military police in the early 1990s. His empire fell and his entire menagerie was rehoused in zoos across South America—except for the hippos. Transporting an animal that weighs up to four and a half tons is an all too hefty challenge for even the most ardent wannabe hippo owner. So for the next two decades, the hippos wallowed in their pond as their master's massive ranch was looted to the point of dereliction, and then transformed by the government into the unlikely combination of Pablo Escobar theme park (open to Escobar fans and complete with waterslides) and high-security prison (open to Escobar wannabes and without waterslides).

All the while, the hippos flourished. Carlos said their numbers were doubling every five years and there may now be more than sixty hippos. I only counted about twenty of them in El Viejo's pond. The rest, Carlos told me, had busted through the flimsy barbed-wire fence that surrounded the hacienda and were running riot in rural Colombia.

Hippos are highly territorial, so as soon as one of El Viejo's sons reaches sexual maturity, the Old Man chases him out of the family pond, away from his harem. The surrounding Magdalena valley is laced with mighty rivers that act as hippo superhighways, allowing these horny young males to travel hundreds of miles into the Colombian countryside. In Africa this would be perfectly normal; young males leave the pod and head off on their own in search of some hippo love. But in Colombia there are no hippos out there, and these frustrated Don Juans are causing havoc.

Carlos took me to meet one such lovelorn hippo, a large adolescent male languishing in a pool just a few hundred feet from a village school. As I approached to get a closer look, the hippo opened his mouth wide in a threatening gape and bellowed aggressively. While I was trying to recall Bill Barklow's hippo language lessons, the hippo started barreling towards me at such speed that large waves rippled out from around his body. The meaning of the call was abundantly clear: intruders were not welcome. I wasn't surprised to hear the kids from the school no longer bathe in the pond, and one boy told me that in the previous week, his grandmother had been chased by this lovesick beast until she nearly collapsed.

Not everyone was scared though. The BBC reported a boy telling the local paper: "My father has captured three. It is nice because you have a little animal at home. They have a very slippery skin, you pour water and they produce a kind of slime, you touch them and it's like soap."

Carlos believed that Colombians are more vulnerable to hippo attacks than Africans because they have grown accustomed to the image of hippos as cuddly animals, the star of Disney movies. "Everyone thinks they are cute animals because they look chubby," he said. "But they're not." Carlos regarded Colombia's hippo population as "a ticking time bomb."

It was not just human safety that worried Carlos. Hippos have the power to radically reshape their environment, and the effect these eco-engineers might have on the local flora and fauna was what worried Carlos most.

IN AFRICA, THE HIPPOS' LIFE is tough. They have to survive a severe dry season. Their pools dry up, food becomes scarce and they must protect their young from hungry crocodiles and hyenas. There are no such challenges for Escobar's ex-menagerie. Steamy Colombia receives year round rain. The hippos have all the grass they can eat, plenty of shallow pools to wallow in during the day, no predators and comparatively little competition from other hippos. Living the good life is changing their behavior. In Africa hippos usually become sexually active between the ages of seven and eleven, but Escobar's hippos are breeding as young as three, and females are reported to be giving birth to a calf every year instead of every two years, as they do in Africa. As Carlos put it, the Magdalena valley is "a hippo paradise."

The normal solution for dealing with an invasive species that's running amok and threatening native species is to extinguish it, like those American beavers we met earlier that were bullying their European cousins to extinction. Governments around the world have gone to elaborate lengths to eradicate rats, ants, mussels and a host of other alien invaders. The government of Guam was praised for its ingenuity for parachuting a fleet of commando cats onto an island to eradicate unwelcome brown tree snakes. And the government of the Galápagos used so-called Judas goats (a reference to Christ's betrayer) to lure invasive goats out into the open, where they were shot by snipers in helicopters. The US government has tried everything from itching powder to poison pellets in a futile attempt to annihilate starlings, which a

New York pharmacist with a misplaced dream of introducing every bird mentioned by Shakespeare into America released into Central Park. Since 1890 his original sixty birds have swelled into a menace numbering in the tens of millions.

When the Colombian military shot their first rogue hippo, however, there was public outcry. The story reverberated through the international media. For a country eager to turn the page on its bloody past, shooting the star of Disney movies was a PR disaster. The eradication program was shelved.

Carlos was the man given the onerous task of implementing Plan B: a radical castration program. Neutering one of nature's deadliest animals is a truly demented idea, but Carlos—clearly a man with serious *cojones*—gave it a shot. Thus far, the Colombian vet has managed to castrate one solitary male that was discovered terrorizing fishermen 155 miles downriver from the hacienda.

For a start, hippos are tricky to anesthetize. They require a large but precise dosage—because of their high fat content (fat absorbs the drug), they are easy to overdose. Administering the drug was also a challenge. Several darts bounced off the hippo's thick hide, and those that pierced it only served to make the famously cantankerous creature even more irate. "We had decoy cowboys to distract the animal's attention, but he still chased us a couple of times," Carlos said.

Once he had got the animal to sleep—it took five darts—his problems were by no means over. These massive mammals are prone to overheating when out of the water, and since underwater surgery wasn't an option, Carlos needed to work fast—as fast as he could, that is. Hippos have internal gonads, hidden beneath several inches of skin and fat (much like their cousins, the whales). To complicate matters further, hippo testicles have been described as "highly mobile" due to their habit of wandering about, especially when under threat. Their position can vary by as much as fourteen inches. Carlos told me he blunted three scalpels during the procedure. It took him a couple of hours to locate these moving targets—even though they were the size of melons—and another hour to sew the bull back up afterwards. In all, the operation took six hours. "A normal castration takes thirty minutes," Carlos said, "but hippos are difficult in every way possible."

Carlos had been loaned an old Russian helicopter to airlift the neutered male back to Hacienda Nápoles. Napolitano, as the hippo was christened, now lives in the big pond. The enormous eunuch is apparently no longer a bother to El Viejo, making the lengthy ordeal an unlikely triumph.

Unfortunately, the operation cost the Colombian government over $150,000—a prohibitive price tag for a developing country with more than its fair share of money-draining problems. Carlos thought there was little chance they would do many more hippo castrations. It looked like the hippos are in Colombia to stay. If they carry on thriving in genetic isolation, they will, in time, become a new subspecies of Colombian hippo, perhaps with the name *Hippopotamus amphibius escobarus*—a most unexpected legacy of a billionaire drug lord as well as a fascinating glimpse into the kind of random event that causes one species to divide into two.

CHAPTER 10

MOOSE

Species *Alces alces*

The Germans call this Beast Ellend, which in their
language signifieth miserable or wretched . . .
—Edward Topsell, *The History of Four-Footed Beasts*, 1607

Evolution doesn't care about human aesthetics—just ask the naked mole rat, an animal that bears an uncanny resemblance to a wrinkly pink penis, with teeth. Its long and winding road creates the most practical, if not pretty, solutions to survival. The moose, whose common American name translates from Algonquian as "twig eater," needs to be able to travel long distances in deep snow in order to sniff out its woody dinner. This largest species of the deer family has evolved to survive the most inhospitable sub-Arctic climes—from North America through to Europe and Asia—on the most miserable of diets. Evolution's solution to this ecosystem challenge is decidedly goofy looking. With its spindly, stiltlike legs, hunched back and long, droopy muzzle, the moose has a certain long-suffering, lugubrious look that invites human misinterpretation. Indeed, so mournful was the moose's appearance, wrote naturalist Edward Topsell, that it had the power to spread its sadness by being eaten. "The flesh," he warned, "engendereth melancholy."

Topsell attributed the moose's melancholy, quite randomly, to chronic epilepsy. "It is in a most miserable and wretched case, for

pag.246. Book 1 Vol.2.   *of animals* 7   *Plat*

The Elk

The Elk falling down in an Epileptick fit being pursu'd by ẏ Huntsmen.

A *Compleat History of Druggs* (1737) features an elk having an epileptic fit. The afflicted animal was said to self-medicate by sticking its cloven foot in its left ear. This too became a standard cure for "the falling sickness" in humans, which is just what every fitting person needs—a dirty great hoof shoved into their lughole.

every day throughout the year it hath the falling sickness," he wrote. This spurious affliction was especially unfortunate for the moose on account of the long-legged creature's allegedly having been denied any knees. They have no "joints in their legs to bend," which led to some rather undignified consequences. "When they are down on the ground, they can never rise again."

Their spurious lack of leg hinges was first described by a man unlikely to have been taken to task for telling tall tales—the great Roman emperor Julius Caesar.

Caesar had encountered moose in the great Hercynian forest. *Alces*, as the Romans called them, were widespread throughout Europe at this time. "They have legs without joints and ligatures," the emperor wrote in his *Commentarii de Bello Gallico*. "Trees serve them as beds. They lean themselves against them, and thus reclining only slightly they take their rest."

Moose actually have rather supple knees—far more so than any other member of the deer family. Their powerful legs can kick in all directions, even sideways.

The Romans' response to the moose's nonexistent disability was to round them up and throw them in the Colosseum. Among the five thousand exotic beasts set to bloody, fatal combat in AD 244 were sixty lions, thirty-two elephants, thirty leopards, twenty zebras, ten moose and one hippopotamus. The bookies must have had a field day.

Back in the Middle Ages, moose roamed all across North America, eastern Asia and Europe, as far south as France, Switzerland and Germany, and were known by a confusing variety of names, including *alg, elch, hirvi, tarandos* and *javorszarvas*. The twelfth-century *Book of Beasts*, written in Latin, refers enigmatically to an "antelope" that historians now believe was most likely a moose.

This "antelope" was praised for its "incomparable celerity. So much so that no hunter can ever get near it"—a statement the stuffed moose heads hanging over many a fireplace would solemnly disagree with. While they may not be able to outrun a bullet or an arrow, moose are indeed swift; faster even than a greyhound and capable of clocking a top speed of just over thirty-five miles per hour.

A combination of their speed, especially in deep snow, and surprising susceptibility to being tamed led the moose to a brief if somewhat unlikely career in the seventeenth century as royal postmen. According to the Scottish naturalist Sir William Jardine, the court of King Charles IX of Sweden employed moose to pull the sledges of his couriers. The monarch even considered founding a moose cavalry, which would have won points for novelty value on the fields of battle, if nothing else.

SUCH UNMITIGATED MISFORTUNE would be enough to turn an animal to drink, and the moose has indeed earned its reputation for being the animal kingdom's number one alcoholic.

September is a busy month for the Swedish police force. It is the season of falling fruit, and for police officer Albin Naverberg, that means drunk moose on the loose. "Just as humans like wine, the moose like fermented fruits," he told me as we drove across Stockholm to investigate the latest of their rowdy misdemeanors.

Sweden is home to around 400,000 moose, which, like the rest of Europe, the Swedes prefer to call elk, though they are the same species, *Alces alces*. (This causes no end of confusion since Americans have decided to call an entirely different species of deer "elk.") Sweden's elk/moose spend most of the year hiding away deep in the forest, far away from the country's enthusiastic hunters, but in autumn they appear to undergo an alcohol-induced Jekyll-and-Hyde transformation. Gangs of marauding moose invade towns and villages, terrorizing the locals like the original stag party. "It's a big problem at this time of year," Albin said. "We get around fifty moose calls every autumn in Stockholm."

Albin and I arrived at a small farm on the outskirts of the capital. It was a slice of bucolic heaven—a rustic wooden cottage nestled in a modest-sized orchard. The heady scent of overripe fruit filled the air. Albin picked up an apple and handed it to me. "This is what they've come for," he said, "but they don't eat the green ones. They only pick out the soft brown ones that have started to ferment."

The homeowners reported that a mother moose and her calf had staked out their garden for a couple of days, gorging themselves on fallen fruit. On the third day, they came out to find the calf dead and its mother absconded.

"They stay in people's gardens for long periods—one or two weeks," Albin said, "and they won't leave the fermented fruit. They think that it is theirs. They get angry if they think anyone is trying to take their fruit away. They can be violent."

You don't mess with a moose. While they may have been unfairly pitted against a pride of lions in the Colosseum, make no mistake: the moose is a badass beast when under threat. The biggest bulls weigh close to a ton, stand over six feet tall and sport a pair of antlers big enough to swing a small hammock between. They use those huge antlers for dueling other males during the rut, or mating season. For us humans it's the legs you've got to watch out for. With the force of four jackhammers, these nimble-kneed ninjas can provide a pummeling worthy of Mike Tyson. The Canadian biologist Jessy Coltrane's advice: "Assume every moose is a serial killer standing in the middle of the trail with a loaded gun." In the state of Alaska, moose are said to attack humans more frequently than bears do (although hard data on this is scant).

According to Albin, moose are most dangerous when drunk. In recent years, mobs of boozed-up moose have terrorized a group of Norwegian hikers and laid siege to a care home of Swedish pensioners (armed police had to chase the horned hoodlums away). In one particularly curious case, a man was sent to jail for murdering his wife, a crime that later turned out to have been committed by a moose that had apparently overindulged on fermented apples.

Moose are not the only members of the deer family accused of druggy delinquency. Reindeer have long been rumored to seek out hallucinogenic fly agaric toadstools, which are alleged to make them "behave in a drunken fashion, running about aimlessly and making strange noises." It's hard to imagine how such sloppy behavior could create an evolutionary advantage. Perhaps, if the rumor is true, some of the other toxins in this famously poisonous red-and-white mushroom help kill off the internal parasites that these ruminants are especially prone to.

The indigenous Sami shamans of Scandinavia are said to use the toadstool's trippy powers in their ceremonies, with their preferred method of intake being a sip of the urine of a reindeer that has been indulging on the fungi. The toadstool's most toxic chemicals are broken down by the reindeer's intestinal system, whereas the psychoactive elements remain intact, making it a "safe" (if fairly unappetizing) means for getting a fly agaric fix.

THE MOOSE'S APPARENT LACK of inhibition is much beloved among the Swedish press. "Swede Shocked by Backyard Elk 'Threesome'" went one headline in the Local newspaper. In the article, a thirty-four-year-old marketing manager called Peter Lundgren was quoted complaining that the elk "were eating apples and then suddenly they assumed the position." He had witnessed a young male moose mounting an older female that was otherwise orally engaged with the rear end of another young male.

But what may be shocking behavior by human marketing manager standards is not necessarily exceptional for elk. Pär Grängstedt, a Swedish zoologist, told the paper that though it was "extremely rare" to see moose mating in a residential area, it was not unusual for them to have sex while others look on. "Usually there are several males in the area competing to be with the in-heat female. The strongest one usually wins, and the others are left to watch," said the moose sex consultant. "It's quite normal behavior."

This public moose-shaming shares moralistic parallels with the medieval bestiaries that employed animals as a means of teaching important moral lessons. Ironically, the moose was chosen by the medieval scribes to lecture people about the evils of alcohol.

The Book of Beasts explained that the moose "has long horns shaped like a saw" which "can cut down very big trees and fell them to the ground." Moose are not known for their lumberjack skills, though the bulls do have a habit of vigorously rubbing their antlers against trees to remove their soft velvet coating just before the rut. The bestiary's author did all he could to find a religious allegory in this deer behavior, relating the elk's two antlers to the two testaments of the Bible, which,

The "antelope" featured in the Northumberland bestiary (1250–60) is depicted getting his comeuppance (and death at the hands of a hunter) after getting caught up with the corrupting "shrub Booze."

he said, can be used to ѕaw off "all fleshly vices," including "drunkenness and lust." Beware "the shrub Booze," the author warned. Its long
branches can entangle the elk's antlers and cause this otherwise swift
creature to become trapped, leading to death at the hands of a hunter.

Beware indeed. In a strange case of life imitating art, Albin told me
that drunk moose frequently have to be rescued from all manner of
entanglements. "I have seen elk that have fallen down hills and end up
hanging from trees," he said. "They often get caught in football nets
and washing lines." These usually do not result in a loss of life, but
there is a significant loss of dignity for all involved.

In Anchorage, Alaska, one local moose, nicknamed Buzzwinkle,
could regularly be found staggering around town trailing long trains
of Christmas lights in his antlers after a big night out hitting the apples. In Sweden, one bedraggled individual achieved international
fame after being caught with its gangly frame entwined among the
branches of "the shrub booze." A snapshot that seemed to capture
its especially humiliating hangover ended up on CNN and went viral,
taking moose-shaming to a new level.

But just because these big gangly beasts with eyes—that can naturally point in opposite directions—look like they're wasted, doesn't
mean they actually are.

MOOSE AREN'T THE ONLY ANIMALS that have been accused of getting plastered on fermented fruit. Newspapers abound with stories of
all types of inebriated beasts from pissed parrots falling out of trees
in northern Australia to orangutans getting blitzed in Borneo on the
sickly liquor produced by overripe durian. There was even a report
of a badger in Germany disrupting traffic by lurching around a main
road after allegedly overindulging on alcoholic cherries. Most of these
stories are purely anecdotal, and about as reliable as the word of a
drunk moose.

African elephants have long been alleged to get wasted on the fermented fruit of the marula tree, which according to an old hunting
bible from 1875 makes them behave like teenagers hitting the town
center on a Saturday night. They "become quite tipsy, staggering

Modern morality fables as told by the popular press: this unfortunate
moose found international fame after becoming entangled in an apple tree
in Gothenburg, Sweden, in 2011. It was allegedly drunk on fermented fruit
and had to be rescued by firemen (who were unable to salvage the animal's
self-respect).

about, playing huge antics, screaming so as to be heard miles off, and
not seldom having tremendous fights."

A natural history documentary entitled *Animals Are Beautiful People*
achieved notoriety in 1974 for capturing the drunken antics of elephants,
ostriches and scores of other animals on camera. The film absurdly an-
thropomorphized its subjects as it showed scene after scene of them
getting loaded on marula and then reeling around with hooded eyes
and wobbly legs, all set to a Benny Hill soundtrack. The footage was
sufficiently compelling to have found a second life on YouTube, where
the video has been watched by more than two million people.

The first person to delve into the truth of this story was the legendary psychopharmacologist Ronald K. Siegel. As associate professor
at the University of California, Los Angeles, he spent a career experimenting with the effects of alcohol and drugs—mostly on human volunteers he called his "psychonauts," but also making the occasional

foray into the wider animal kingdom. He has given monkeys cocaine chewing gum and claims to have taught pigeons "how to tell us what they were seeing while under the influence of LSD." To which the somewhat mundane answer was: blue triangles.

In 1984, Siegel undertook a significantly more treacherous study into what happens when you give a group of captive elephants "with no history of alcohol use" access to unlimited alcohol. He discovered they were more than happy to drink the equivalent of thirty-five cans of beer a day, enough to partake in "inappropriate behaviors" such as wrapping their trunks around themselves, leaning against things with their eyes closed and dropping their trunkhold on each others' tails, which Siegel described as "the trained elephants" version of walking in a straight line.

Playing bartender to a herd of elephants was not without its risks. A large bull named Congo chased Siegel's jeep after the professor tried to cut off his beer, attacking Siegel with the empty barrel. On another occasion, Siegel had to break up a fight between Congo and a sober rhino that happened to wander into the elephant's favorite waterhole at the wrong moment. "I knew a life-threatening clash was imminent." Siegel decided to drive his jeep between the two animals, only narrowly avoiding becoming part of the fight himself. "I should have known better," he wrote later.

Siegel's rather whimsical conclusion from his elaborate drunken circus was that the elephants did indeed drink until drunk, doing so perhaps to forget "the environmental stress" of their ever-shrinking homeland and the competition for food. But just because elephants can get drunk when supplied with a steady stream of alcohol, doesn't mean they are getting similarly wasted on fermented fruits in the wild.

While attending a physiology conference in South Africa, a group of British biologists from Bristol University took a more sober scientific approach to the investigation than had Siegel, and instead of dosing elephants irresponsibly with unlimited booze, they used statistics to find their answer. They created various mathematical models based on an average elephant's weight and the alcohol content of the marula fruit and calculated that an elephant would have to eat marula at 400 percent its normal feeding rate to get looped. "These models were highly biased in favour of inebriation," the researchers said, "but even so failed to show that elephants can ordinarily become drunk."

The biologists branded the marula story as yet another zoological myth driven by our desire to humanize animals. The boozed-up stars of *Animals Are Beautiful People* had, it seems, been injected with veterinary anaesthetic in order to elicit their woozy behavior. "People just want to believe in drunken elephants" was the researchers' final conclusion.

The same would seem to be true of the moose. One Swedish professor told me that there had never been a test confirming high blood alcohol in an elk. "At this point I think the idea rather reflects our Nordic-Germanic problematic relationship to alcohol."

Rick Sinnott, the Canadian biologist who spent many years trailing the infamous Buzzwinkle around Anchorage, told me that he suspects a more likely explanation is that the moose are suffering from apple acidosis, brought on by consuming an unnatural amount of sugar-rich food. This causes a buildup of lactic acid in their gut, a condition that can result in symptoms that include dilated pupils, a struggle to stand and severe depression—all of which sounds remarkably like the early naturalists' portraits of the moose. It seems the animal they were describing was neither alcoholic nor melancholic, but instead suffering from a case of acute indigestion.

Which is not to say no moose has ever been blotto. Indeed, there appears to have been at least one—a pet belonging to the sixteenth-century Danish astronomer Tycho Brahe, whose accurate preltelescopic observations laid the foundations of modern astronomy.

Tycho was an unusual character. He lost his nose as a student in a duel over maths and was forced thereafter to wear a false one fashioned from brass. He built his very own castle, complete with underground laboratory, on the island of Hven and invited the great and the good to join him there for lavish parties. There they were entertained by a psychic dwarf called Jepp and Tycho's pet moose, which, according to the astronomer's diaries, was an exceedingly good sport: "It prospers, runs about, dances and is of good cheer . . . just like a dog."

Though he was clearly very fond of his pet, Tycho agreed to gift it to his patron in an effort to increase the astronomer's standing in society. It died en route, at a castle in Landskrona, where it was alleged to have consumed a quantity of beer before taking a fatal tumble down the stairs.

Perhaps this was the only case of a genuinely drunken moose. But it's worth considering that a sober moose might have had some trouble going down the stairs as well.

GIVEN THE MOOSE'S REPUTATION for drunken delinquency, it is a little surprising that in the late 1780s, a rather dilapidated dead moose became the unlikely defender of the newly independent American colonies' honor.

The Old World's deep suspicion of the New was crystallized into a neat scientific doctrine by the imperious George-Louis Leclerc, Comte de Buffon, who outlined the "Theory of American Degeneracy" in his *Histoire Naturelle*. "In America," he claimed, "animated Nature is weaker, less active." Not only were there fewer species to be found in the New World, but "all the animals are much smaller than those of the Old Continent." Of those found on both continents, the ones of the New World were more "degenerate." And America boasted no showstopping giants: "No American animal can be compared with the elephant, the rhinoceros, the hippopotamus, the dromedary, the camelopard [giraffe], the buffalo, the lion, the tiger."

Buffon had never actually set foot in the New World but, as usual, that was no impediment to his theorizing. His knowledge of the diminutive nature of American animals was mostly gleaned from stuffed specimens and travelers' tales. He had, however, developed an idiosyncratic system of accountability: if he read the same "fact" in at least fourteen travelers' tales then, according to his peculiar probability calculations, he was able to be certain of its "moral certitude."

Buffon claimed the North American continent had only recently lifted itself up out of the sea and was mostly swamp; unlike Europe, it was still drying out. That's why American animals and plants were smaller, weaker and less diverse. The only creatures that grew to any size were insects and reptiles, "all the animals which wallow in the mire, whose blood is watery, and which multiply in corruption, are larger and more numerous in the low, moist, and marshy lands of the New Continent."

The dogs of America were both diminutive and "absolutely dumb." And in a uniquely French insult, the Comte proclaimed that the lamb raised in the New World was "less juicy."

The Native American peoples were, according to Buffon, equally degenerate, weak of mind with hairless bodies, a lack of "ardour" and "small and feeble" genitals. Although he stopped short of actually saying it, the suggestion was that any European going to America would suffer the same diminishing effects, since any animals transported to America "shrink and diminish under a niggardly sky and an unprolific land."

Buffon's theory was a massive slap in the face for the developing nation. The prospect of giant insects and shriveling genitals wasn't exactly a great advert for a country desperately trying (in those days) to appeal to immigrants. The Comte was the most famous naturalist of his time, a leading light of the Enlightenment whose encyclopedia was an international bestseller. As a result, his theory of American degeneracy caught on like wildfire, providing Europeans with a handy scientific justification for believing that their continent remained superior to this Brave New World.

In the name of American virility, something had to be done. Enter Thomas Jefferson, future president of the United States, who somehow found time in between authoring the Declaration of Independence, holding office as governor of Virginia, and serving as ambassador in Paris, to take on the Comte's withering claims. Jefferson loved nature almost as much as he loved politics, so he was perfectly placed to retaliate against the belittling of his beloved nation with hard evidence of American greatness.

Jefferson wrote to his fellow Founding Fathers, urging them to head out into the wild with a pair of calipers and start measuring American animals so that they might counter the French naturalist with their own set of diplomatic data. The political men responded to this task

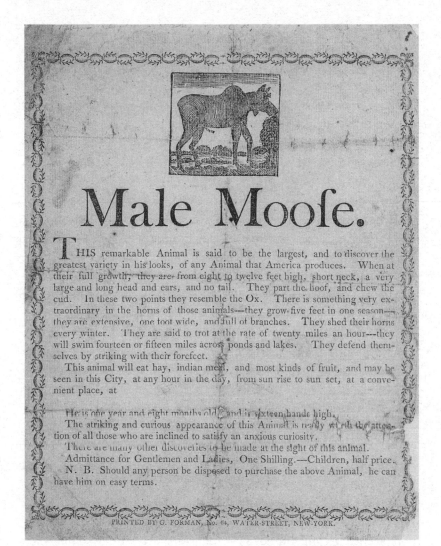

This 1778 advertisement boasting the greatness of the male moose was probably an attempt to crush the Comte de Buffon's withering "Theory of American Degeneracy," which accused New World creatures of being weak and feeble. It claims that "this remarkable animal" might be as much as twelve feet high—a tall tale indeed.

with great gusto. James Madison sent Jefferson a lengthy missive in which he first debated the merits of different forms of representative government, and followed it up with an extraordinarily precise description of a local Virginian weasel, measuring each and every part in three dimensions right down to the "distance between the anus and the vulva." Madison concluded that the weasel's statistics "certainly contradicts his [Buffon's] assertion that, of the animals common to the two continents, those of the new are in every instance smaller than those of the old."

While Jefferson was serving as the US ambassador to France, he received an invitation to the Comte's summer home in Paris. It must have been a tense soirée. After initially trying to ignore each other in the garden, they collided in the library. Before Jefferson could dazzle Buffon with his comparative weasel stats, Buffon plopped a massive manuscript in front of him, the latest version of his encyclopedia, and said, "When Mr Jefferson shall have read this, he will be perfectly satisfied I am right." The evening dissolved into an argument about moose.

Buffon said he was "absolutely unacquainted" with the American moose and thought it was simply a miscategorized reindeer. Jefferson told the Comte, somewhat rashly, "that the reindeer could walk under the belly of our moose."

Finally, Buffon threw down a giant gauntlet: if Jefferson could present him with a moose "with horns one foot long," he would retract his theory of American degeneracy in the next volume of his natural history tome.

Jefferson had known for some time that moose were his trump card. He'd already circulated a sixteen-question survey to gather information on such pertinent issues as "Do they make a rattling sound when they run?" He had also begged his political allies to shoot, stuff and post him their largest specimen, "seven to ten feet tall" with "horns of a very extraordinary size"—the very definition of a tall order.

One General John Sullivan had emerged as Jefferson's main moose man. After the encounter with the Comte, Jefferson wrote to General Sullivan with increased urgency. "The readiness with which you undertook to endeavour to get for me the skin, the skeleton and the

horns of the moose," he wrote from Paris on January 7, 1786, "embold-
ens me to renew my application to you for those objects, which would
be an acquisition here, *more precious than you can imagine.*"

Jefferson went into minute detail about how the animal should be
prepared in order to preserve its greatness. "Leave also the bones of
the head in the skin with the horns on," he wrote in his lengthy set of
instructions to Sullivan, "so that by sewing up the neck and belly of
the skin we should have the true form and size of the animal."

When Sullivan finally procured a seven-foot moose for Jefferson,
it had already been on the road for fourteen days. The general was
no taxidermist and he struggled to fix the toll taken on the animal's
carcass, which was in "a state of putrefaction." The moose's magnificent antlers were also crucially absent. So Sullivan was forced to send
the corrupt creature with a smaller, surrogate pair. "'They are not the
horns of this moose," he confessed to Jefferson, "but may be fixed on
at pleasure."

After yet more mishaps, including almost getting lost at a ship dock,
the moose finally arrived in Paris in October 1787. By then it was in a
decidedly forlorn state: misshapen, virtually bald and minus its massive antlers. Jefferson was nevertheless optimistic and forwarded the
animal to Buffon with a note excusing the unusually pathetic appearance of this great beast, in particular its horns, which "are remarkably
small." He added, "I have certainly seen of them which would have
weighed five or six times as much."

Jefferson wrote in his journal that Buffon, despite the moose's miserable appearance, "promised in his next volume to set these things
right."

Alas, the timing was tragically off. The Comte died shortly afterwards, in 1788, with no word of a retraction.

Undeterred, Jefferson went on to develop his argument and published a comprehensive table of comparative New and Old World animal statistics in his *Notes on the State of Virginia.* The book became a
bestseller in its own right, and the degeneracy theory fell into a state
of dissolution.

CHAPTER II

PANDA

Species *Ailuropoda melanoleuca*

> *Pandas are bad at sex and picky about food . . . Females are*
> *in heat only for a few days a year; males have the savoir faire*
> *of pimply prom dates. These genetic misfits might have*
> *died out long ago, had they not been so adorable.*
> —The Economist, 2014

The panda is considered a joke of a bear, universally adored for its cute comic-strip looks but ridiculed for an apathetic approach to sex and deviant vegetarian diet. Even respected naturalists have declared that "it's not a strong species" and dared call for its demise. No other animal has been forced to justify its existence as much as the panda, which, without our help, would surely join the dinosaurs and the dodo on evolution's scrap heap.

Or would it?

This image of the pathetic panda is a very modern myth. There are, in fact, two pandas. The one that lives in zoos and gets celebrity treatment in the media is a cartoon of our own creation, a benign bungling comedy act that does indeed need human help in order to exist. The other panda is a splendid survivor that's been around for some 18 million years—three times as long as us hominims—and perfectly adapted to its admittedly eccentric lifestyle. This wild panda is a secret stud, fond of threesomes and rough sex, with a taste for flesh and a fearsome bite. But it inhabits an impenetrable forest in a cryptic country, which has allowed an imposter to take center stage and ensured that one of the most recognizable animal brands is nothing but a fraud.

FOR AN ANIMAL as phenomenally famous as the panda, it is surprisingly new to science. As little as 150 years ago, it languished in virtual anonymity, with remarkably few references to its existence even within its native China. That changed in 1869, when a French missionary set off the rhetorical rocket that would launch the panda into global superstardom.

Father Armand David loved nature almost as much as he loved God. "It is unbelievable that the Creator could have placed so many diverse organisms on the earth," he wrote in his diaries, "only to permit man, his masterpiece, to destroy them forever." The good father indulged both of his devotions by exploring China—a country where he could convert plenty of "infidels" to his particular flavor of Catholicism while searching for new species to send to the natural history museum in Paris. We do not know how many Chinese souls he managed to save, but the keen-eyed missionary certainly had an impressive strike rate in uncovering new species: a hundred insects, sixty-five birds, sixty mammals, fifty-two rhododendrons and one frog ("that barks like a dog") were all first discovered by his hand. His most enduring legacy might have been giving the world gerbils—an unquestionably prolific species—if it hadn't been for a chance encounter that guaranteed his place in zoological history.

One afternoon, while taking tea at a hunter's house in the mountains of Szechuan, Father David stumbled upon the pelt of "a most excellent black and white bear" that he believed might prove "an interesting novelty to science." A few days later his "Christian hunters" brought him back a specimen that the priest noted "does not look very fierce" and had a stomach "full of leaves." He named it *Ursus melanoleucus* (literally "black and white bear") and sent its skin to Alphonse Milne-Edwards, director of the National Natural History Museum in Paris, for classification.

The animal was unknown to science, and quite a find, but Milne-Edwards was not convinced by the priest's assessment. The teeth and unusually hairy underpaws resembled those of another bamboo-munching mammal recently discovered in the same Chinese mountains, a relative of the raccoon that had been named the red panda (*Ailurus fulgens*). So Milne-Edwards declared that this black-and-white bear join the red panda in its own family, the *Ailurapoda* (meaning "panda foot"), quite distinct from bears.

So began more than a century of taxonomic toing and froing over the status of the anomalous giant panda, the arguments ranging from the scientific to the highly subjective. The molecular biologists tussled over mitochondrial DNA and blood proteins; others, who should have perhaps known better, took a more instinctive approach—including esteemed biologist and conservationist George Schaller, who, despite overwhelming evidence to the contrary, continued to insist that "even though the giant panda is most closely related to the bears, I think that it is not *just* a bear." Schaller felt in his gut that lumping the panda in with the other *Ursus* destroyed its uniqueness. "The panda is a panda" was his professional opinion on the matter. "Just as I hope there is a yeti but that it will never be found, so I would like the panda to retain this minor mystery."

Alas for Schaller, the geneticists poured irrefutable ancestral evidence all over his dreams. When the panda's genome was sequenced in 2009 it unequivocally exposed the animal's bear necessities. Father David had been right: the panda was not a panda but the most ancient of all the bears, an early offshoot from the *Ursus* lineage that diverged from the rest some twenty million years ago. But the panda name stuck, promoting its mythical otherness.

Once you scratch beyond the surface, the giant panda is not so different from its bear brethren. The panda's quirky reproductive cycle, boasting a brief fertility window and an ability to delay implantation of the fetus, is also found in other bears. Panda babies are born decidedly undercooked—blind, pink and roughly the size of a mole—like all bears, whose tiny newborns are less than one percent of their adult

The bear from this French bestiary (c. 1450) may look as if it is indulging in
a little casual coprophagy, being rather sick or both. It is instead illustrating
the myth that bear cubs are born as unformed lumps that are "licked into
shape" by their mothers.

body size. This fact led early naturalists, such as Claudius Aelianius,
writing in the third century, to the conclusion that a bear "gives birth
to a sort of misshapen lump, with no distinctive form or features" and
the mother then "licks it into the form of a bear"—an imaginative mistake that provided the source of the phrase "lick into shape."

Aelianius had a way with words, which he put to effective if inaccu-
rate use in his descriptions of bear behavior. He observed that bears
like to hibernate without food or water, causing their intestines to "dehydrate and atrophy," which the bear reactivates by eating wild arum.
This, he said, "makes her fart," before she "eats a bunch of ants" and
"enjoys a fine defecation."

Pandas neither hibernate nor self-medicate but, strictly speaking, no
bears do. A handful of species—the black, brown and polar bears—do

enter the long winter sleep known as torpor, during which their body temperature drops, they fast and they refrain from excreting, but it's not considered a true hibernation. How much they "enjoy" their first post-sleep poo is something we will likely never know.

The panda's vegetarian diet may be rather narrow, but it is not so terribly unusual among bears. Most bears, despite being classed as carnivores, are actually opportunistic omnivores, with plants account-ing for at least 75 percent of their diet. The panda has taken this to an extreme and dines almost exclusively on bamboo, which is more than plentiful in their native mountains. Other bears are equally picky about their food. The sloth bear is adapted to eat only termites (and has lost its front teeth to simplify sucking them up with an extra long tongue), and the polar bear has a preference for feasting almost exclu-sively on ringed seals. This dietary specialism doesn't, however, mean that the panda has lost its taste for flesh. When George Schaller was studying pandas in the wild, he would bait the traps with goat meat, as this was a sure-fire way to attract them. I've seen footage of a wild panda chowing down on a dead deer. Panda eats Bambi is decidedly NSFD—not safe for Disney.

It's THE GIANT PANDA's sexual appetite that is the most misconstrued. This modern mythology has been created by a parade of pandas para-chuted into foreign zoos where, thanks to us, they've acted out a sex-ual farce worthy of a 1970s sitcom, making them and their laughable sex lives very famous indeed.

The first of the monochrome bears to grace foreign soil landed in the US just before the start of the Second World War, bringing a dose of much needed joy to Americans weary from years of suffering un-der the Great Depression. First came a roly-poly baby named Su-Lin, meaning "a little bit of something cute." Her anthropomorphic antics and extreme exotica were only amplified by her chaperone—fashion designer and socialite Ruth Harkness. This unlikely female explorer had faced bereavement, bandits and bewildering bureaucracy, not to mention the indignity of traveling by wheelbarrow, in order to pluck the black-eyed ball of fluff from the mountains of China with her own bare hands. The deliciously scandalous story of her panda-snatching, complete with rumors of illicit romance and the advent of a dastardly competitor (who allegedly tried to smuggle a panda out of China before Harkness by dyeing the bear brown), had fed the newspapers for months. So when the explorer and the bear finally stepped off the boat, Su-Lin received a welcome worthy of a movie star.

The Shirley Temple of the animal world did not disappoint. Humans are preprogrammed to want to nurture anything with neotenous, or babylike, features—namely, a big bulging forehead, large, low-set eyes and round protruding cheeks. It's a neurochemical insurance policy to ensure we take good care of our unusually vulnerable babies, themselves a result of our big brains, which require a birth in early development in order to allow our relatively humongous heads to exit the birth canal safely. It's such a deeply hardwired impulse, we will even respond lovingly to inanimate objects that vaguely exhibit these characteristics—like the VW Beetle.

Pandas, with their unique markings and decidedly humanlike way of sitting and eating, could have been genetically engineered as the perfect trigger for this nurturing instinct. They fool our brains into firing off the reward center, the same part that responds so delightfully to sex and drugs. So baby giant pandas are essentially cute crack. The sight of Su-Lin being bottle fed by her human "mother" and acting up like a naughty toddler for the assembled cameras left America in a warm and fuzzy puddle.

Su-Lin was soon followed by Mei-Mei, her "little sister," and finally a potential beau named Mei Lan. Attempts by the Chicago zoo to get them to mate were hampered by the fact that all three were actually male. So as the world watched with bated breath for a budding romance, the two male pandas did nothing but disappoint, and the press reported their every failed move. "Panda Love: Mei-Mei Courts Her Beau and Gets Nowhere at All," shouted a typical headline from *Life* magazine, sowing the first seeds of the sex-shy myth.

A similar fate befell the Bronx Zoo's "breeding pair," which arrived to much fanfare in 1941. Pan-dee and Pan-dah—an early example of the "Boaty McBoatface" rule of never naming anything of any importance via the medium of public contest—were not a little boy and girl but two females. "The external differences that the Bronx officials studied were obviously individual characters rather than true sexual ones," explained the zoologist Desmond Morris helpfully in his 1960s book *Men and Pandas*.

Sexing pandas would prove to be a notoriously difficult art, with countless other cases of mistaken gender leading to further sexual disappointments. The truth didn't help the panda's case much either, since the panda penis is virtually indistinguishable from female genitalia.

When a male and a female panda finally found themselves in a cage together, the results were no more fulfilling.

One of the most influential of these failed liaisons starred a panda called Chi-Chi. The young Chi-Chi arrived at London Zoo in 1958 and found herself the star of a hit TV program that documented her every move for an adoring public, like a panda *Truman Show*. After several years of carefully staged bubble baths and football games with her keeper, the soap opera of Chi-Chi's life demanded a love interest. The only other panda in captivity outside of China was An-An, a resident of Moscow Zoo. So at the height of the Cold War, an unlikely East-West union was planned, much to the glee of the international press corps.

Chi-Chi, however, was less than keen. Once in Moscow, she repeatedly fought with An-An, spurning his clumsy advances. Chi-Chi, it transpired, wasn't attracted to pandas; her desire appeared to be directed towards humans, especially ones in uniform. When the Soviet zookeeper entered her cage, Chi-Chi "raised her tail and rear-end in sexual response," reported Oliver Graham-Jones, the London Zoo's curator of mammals—much to the Russian's "intense embarrassment."

This was not the first time this had happened. "Sexually, Chi-Chi appeared to be somewhat warped," recalled Graham-Jones. She was more attracted to zoo staff and "even complete strangers" than she was to her own kind. It was the 1960s, the era of free love, but after three failed attempts at panda romance, it was clear there would be no sexual fulfilment for the zoo's superstar. The Zoological Society

A romantic scene from 1959—Chi-Chi with her true love, London
zookeeper Alan Kent, as he tenderly feeds her bamboo from his own
mouth. A decade later, the zoo was forced to admit their famous female
panda had sexually imprinted on the wrong species.

of London was eventually forced to issue a statement declaring: "Chi-
Chi's long isolation from other pandas has 'imprinted' her sexually on
human beings."

Chi-Chi and An-An's doomed romance was no one-off. Shortly
after it fizzled out, another high-profile pair of pandas, Hsing-Hsing
and Ling-Ling, arrived at the National Zoo in Washington with an important mission—to save their entire species. Chi-Chi had inspired the
World Wildlife Fund's famous logo, and the growing panda conservation movement was exceptionally focused on breeding the bears in
captivity as part of their survival plan. Unfortunately, the bears had
not read the conservation memo and their high-profile romance was
confined to yet more underproductive antics. Hsing-Hsing exhibited

an "orientation problem," which resulted in Ling-Ling's ear flap, wrist and right foot being the targets of his amorous attempts. But now these weren't just personal failures; they spoke to the fate of the species. The press reacted with much ribald. The public responded by sending the pandas a waterbed.

For the next two decades, top zoologists delved into the biology of the Washington pandas in search of answers, while the press and public bayed for babies. Ling-Ling's estrus was discovered to last less than two days every year—a frustratingly narrow fertility window that took much of the blame. Then, when the capital couple finally did manage to get it on, their offspring did not survive more than a few days. One cub perished after Ling-Ling sat on it.

Each of these very public dramas added to the perception that pandas are simply not built to procreate *or* parent, and have somehow been denied the fundamental instincts required for survival. The calls to take human control of the situation and find some way to force pandas to reproduce in captivity became increasingly urgent.

Breeding wild animals in captivity is rarely easy. And it takes just a little common sense to see why: a concrete enclosure is not a sexy place for a wild animal. Their desire to procreate has to be stimulated by a range of behavioral and environmental cues—the animal equivalent of a nice glass of wine and a bit of Barry White. Often zoos have no idea what is required to get their animals in the mood. The white rhino, for example, was proving impossible to breed in captivity, because zookeepers were simply bunging a male and female in a pen together and hoping for the best. This didn't take into account that rhinos are herd animals, and for a male to get the horn, as it were, he has to flirt with *several* females before choosing the lucky lady. In giant pandas, it is the reverse—the female is the picky one.

In the 1980s, George Schaller was the first person to discover that these notoriously solitary creatures are not loners when it comes to sex. Crawling through bamboo thickets in the fog and snow, he observed complex mating rituals in which lone females climbed trees and moaned like Chewbacca as several males fought for her attention below. The winning male would take advantage of his prize by having sex over forty times in a single afternoon. The semen of the giant panda is also said to contain "prodigious amounts of high quality spermatozoa," ten to a hundred times more than a human male. There is no denying that these are virile animals.

Panda sex itself is a rough-and-tumble affair with plenty of biting and barking. The male probably learns the right amount of submissive-dominant behavior from playing with his mother and observing her in the act; baby pandas stay under their mother's care for up to three years. This gives them the opportunity to witness at least one breeding season and learn the ins and outs of the female panda's preferred *Kama Sutra*.

Pandas occupy sizeable home territories of around 2.5 to 4 square miles, and they sniff out sex parties by leaving scented status updates advertising their identity, sex, age and fertility on specially designated trees—the panda equivalent of Tinder. When a female comes into season, she arouses male interest by rubbing her anal glands at the base of one of these communal message boards. Her smelly signal attracts males from far and wide, who then compete for her affections in a sort of urinary Olympics. Female pandas prefer males that can leave their sexy scent marks the highest up a tree. Scientists have described males adopting a selection of athletic poses—"squat," "legcock" and, most remarkably, "handstand"—in order to squirt their pee as high as possible. Males are also thought to use their own bodies as sexy-scented advertising boards by dabbing urine on their ears, like aftershave, which act as a pair of fluffy beacons broadcasting the bear's availability on the mountain breeze.

Bears are famous for their highly developed sense of smell, so the female's short fertility window is no impediment to reproduction in the wild. Indeed, it could even be an evolutionary adaptation to control population size, precisely because male pandas are so accomplished at procreating, which helps ensure the birth rate never exceeds what the bamboo forests can support. A wild female will give birth to a cub on average every three to five years, which is not an unusual reproductive rate. If they reproduced any more quickly, they would quickly outgrow their habitat.

This wild sex is a world away from being thrown into a concrete enclosure with a random panda. Yet, in the last couple of decades, the Chinese have cracked the captive breeding conundrum to produce bumper crops of baby pandas, which make for images so cute they should come with a public health warning. Dozens of baby pandas lined up in a row have guaranteed these conservation "success" stories get attention around the globe.

In 2005, I flew to China to find out how they were doing it for a documentary I was researching. My destination was the Szechuan capital of Chengdu, home to the most bountiful of China's specially built panda breeding centers. I was expecting the heart of panda country to be a green and pleasant land, but I was instead greeted by a sprawling city of fourteen million people. The first panda I saw was on a packet of cigarettes, which seemed oddly appropriate given the amount of smog that hung in the heavily polluted air. There is a saying in Chengdu that the dogs bark when they see the sun. They were resolutely silent during my entire visit.

My visit to the Chengdu Research Base of Giant Panda Breeding started with a tour from one of the senior scientists working there. As we wandered around the gray concrete buildings, he told me about some of the more imaginative approaches they'd taken to getting pandas in the mood in this astoundingly bland and brutalist environment. Aware that the young male pandas had not received the necessary sex education from their mothers (most were hand-reared by humans from birth), the scientists were experimenting with plonking three-year-old boy pandas in front of a portable TV playing a VHS recording of captive pandas humping away as a kind of coming-of-age present. Given the panda's notoriously poor eyesight, and the difficulty any animal (other than humans) has wrapping its brains around the concept of seeing itself on TV, panda porn seemed unlikely to do anything more than raise a laugh. As did their experiment with using sex toys to stimulate the female pandas. Viagra had also been supplied at other panda breeding centers; one underperforming sixteen-year-old named Zhuang Zhuang ("strong strong") had been given the first experimental dose, but nevertheless failed to live up to his name.

I couldn't understand how such humancentric solutions to the panda problem could have contributed to the extraordinary boom in babies. And it turned out they hadn't. The strategy that was paying baby panda dividends was something significantly less saucy: artificial insemination (AI).

AI has become standard practice for coaxing reproductive success out of all sorts of captive animals. All you need is some viable sperm and a sleepy female.

Sometimes the semen is collected using "digital manipulation," euphemistic science-speak for an animal hand job—a not uncommon procedure in the conservation world. I once witnessed a prizewinning Shire horse being "harvested" for his "liquid gold" (something my eyes will never unsee). The last pure Pinta Island tortoise, a Galápagos native named Lonesome George, had his very own personal "digital manipulator"—an attractive young Swiss zoologist whose task was to bottle as much of this living relic's seed as she could possibly extract. She managed to hone this skill sufficiently to relieve the slow-moving centenarian in under ten minutes, earning her the nickname "George's girlfriend."

Such exploits are not without hazard for the designated manipulator. A senior scientist at a conservation center in Berlin sustained a serious black eye from trying to manually massage a bull elephant's flailing three-foot-long penis to orgasm, which must have taken some explaining down the pub afterwards.

A safer and more dignified solution for the human, if not the animal, is electro-ejaculation. This involves inserting an electric probe up an animal's rectum and cranking up the voltage until the animal climaxes. Dr. Kati Loeffler, an American veterinary scientist who spent several years working at the Chengdu center and the Smithsonian Institution, explained to me that this technique was developed for the intensive breeding of farm animals. It is now standard practice for pandas in China, which are dosed up with trance-inducing ketamine to help mitigate the discomfort associated with being anally raped with an electrified truncheon. According to Loeffler, these factory-farming

The year 2016 saw a bumper crop of baby pandas for the Chengdu breeding center, seen here being wrestled into position for the all-important "cute crack" photo op that will send the story of their "conservation success" worldwide.

techniques were the reason the captive panda population had soared to almost five hundred individuals.

This is not a conservation success story, however. While these black-and-white balls of fluff look like pandas, they don't grow up to *behave* like pandas. Inseminated females frequently give birth to twins, which are kept alive by switching them between their mother and an incubator so both get a chance to suckle her milk and develop physically, with the necessary immune system for survival. Then, at three to four months they are removed from their mother completely so she can return to the captive breeding conveyor belt. The babies are subsequently raised by humans in an unnatural pack, until they get too big and belligerent, at which point they get placed in solitary confinement.

"Cubs in breeding centers and zoos are raised in a human-intensive environment in which opportunities for normal social and behavioral development are severely compromised," Loeffler said. "Young pandas never have a hope of learning to become a normal panda."

When attempts have been made to reintroduce these offspring into the wild—which is the much-heralded purpose of these captive breeding efforts—the results are little better. Recent studies have shown that these so-called solitary animals are quite social, even outside of the breeding season. "In the wild, bears have to be very savvy socially," said Loeffler. "They learn these very sophisticated social negotiations that allow them to work together and share a food resource as well as at mating time." A socially awkward panda risks more than just going hungry or not getting laid.

A young male named Xiang Xiang was the first panda to be reintroduced to the wild from a captive breeding program in China. Loeffler told me it was no accident they chose a male to experiment with. "They need to keep the females because they're the ones that make the golden eggs." She said he did okay for the first few months, "but then came time for breeding season and you have all these males gathering around the estrus female and of course he had no idea what to do—he'd been raised in isolation behind bars—and of course they beat him up so badly that he was nearly killed." He was eventually found dead, savaged to death by wild pandas. To date, of the ten pandas released into the wild, only two remain there.

Releasing captive-born animals into the wild is a bit like chucking a Chihuahua into a pack of wolves. In recent years, one of the Chinese breeding centers has made efforts to better prepare their pandas for wild life. Designated mothers are permitted to stay with their babies in semiwild enclosures. Their human handlers wander around in panda onesies, pushing about taxidermy leopards mounted on wheels in an attempt to teach the babies about predators. The surreal scene makes for a great photo opportunity, but Lü Zhi, professor of conservation at Peking University, has called the reintroduction efforts "as pointless as taking off the pants in order to fart."

Dr. Sarah Bexell, a conservation scientist with a CV that includes successful reintroduction schemes for golden lion tamarins and black-footed ferrets, explained why. "For one thing, there are no longer leopards in panda habitat," she said. More importantly, "Humans cannot teach animals how to be animals. Only their mothers, or other conspecifics"—members of the same species—"can be the teachers." Since these panda moms have been reared by humans, they have little or no experience themselves of life in the wild to transfer to their cubs.

The biggest issue of all is the panda's shrinking habitat, something I noticed on the long drive from Chengdu to the Qinling Mountains, where wild pandas are said to still roam. It took several hours to emerge from the city. We passed miles of sprawling urban detritus, then factories belching cement dust over entire villages like something out of a dystopian sci-fi movie. Once we reached the mountains, we were greeted by monstrous dams. And that was more than a decade ago.

The Chinese government has created more than fifty panda reserves that claim to have witnessed an increase in pandas—so much so that the panda's status was recently downgraded from "endangered" to "vulnerable." But Kati Loeffler isn't convinced. She told me the authorities had zoned these "protected" areas to allow farming, roads and even mining inside them. "We are so arrogant that we think we need to breed pandas and put them back in the wild because they are too stupid to manage on their own," she said. "If we just gave them back their habitat, they would do the job themselves, like any other species. There is nothing terminally wrong with these animals that we need to fix; the only thing we need to fix is to give them their home back."

Sarah Bexell agreed. "I really worry that it is wildlife conservation's way of saying, 'See, guys, we can do this, we are scientists and we're fixing this huge biodiversity crisis problem and we can right the wrongs of our entire species by doing these little projects.' It's a feel-good story. The public feels good and sits back and eats their potato chips in their La-Z-Boys and drives their SUVs, has a five-bedroom house and three kids, and says, 'Great, those scientists are out there fixing this problem for me,'" she said. "But science is not going to save biodiversity; a shift in human behavior is the only thing that is going to save it. I think there needs to be more effort, first and foremost, in getting the human population under control globally, and for people to start contemplating not being such mass consumers."

Paradoxically, the pandas themselves have helped grease the wheels of China's explosive expansion, and the associated environmental cost, with their extraordinary diplomatic skills.

A few years ago, I spent some time with two such political pandas: Tian Tian and Yang Guang, a.k.a. Sweetie and Sunshine. They had arrived in Edinburgh in December 2011 on board their very own private Boeing 777 airliner, complete with a giant panda painted on the side. Flag-waving schoolchildren lined the streets to greet them, bagpipes played and a specially commissioned panda tartan was unveiled in their honor (I'm sure it made the pandas feel very welcome). The whole event was covered by live rolling news. Sweetie's and Sunshine's job was to spread a bit of panda magic at the local zoo's flagging ticket office, with the aim of boosting sales by some 70 percent. But their arrival was also underpinned by other, less publicized economics: a set of trade deals through which Scotland would provide China's burgeoning middle class with factory-farmed salmon among a suite of contracts worth an estimated $3.5 billion. In recent years, globetrotting pandas have accompanied similar arrangements for uranium (with Australia) and seal meat and petroleum (with Canada).

"The panda can be used to seal the deal and signify a bid for a long and prosperous relationship," Dr. Kathleen Buckingham, lead author on a study of "panda diplomacy," told the BBC. "If a panda is given to the country, it does not signify the closing of a deal—they have entrusted an endangered, precious animal to the country; it signifies in some ways a new start to the relationship." China, she said, was interested in having "soft power influence through a global visual seal of approval." The panda seems tailor-made for the part.

Panda diplomacy is not new. Way back in the seventh century, the Tang dynasty presented a pair of live pandas (along with seventy of their skins) to the rulers of Japan. The policy was resuscitated in 1941 when China sent Pan-dee and Pan-dah to the Bronx Zoo as a thank-you gift for American aid during the Second Sino-Japanese War. Chairman Mao was a huge fan of the diplomatic power of these iconic black-and-white bears. During his time in power pandas were dispatched to long-term Communist partners like North Korea and the Soviet Union as well as new political allies. When Nixon brought a close to twenty-five years of antagonism between the United States and the People's Republic with his landmark trip to China in 1972, the sexually underperforming Hsing-Hsing and Ling-Ling were dispatched to Washington and sparked a general "Panda-monium."

In exchange, the White House also presented China with a pair of animal ambassadors. The Americans could have chosen stately bald eagles or powerful grizzlies to represent their country, but instead a couple of shaggy, smelly musk ox were forced to step up. The notoriously belligerent beasts failed to illicit quite the same commotion in China as the pandas had in America. Milton, the male, was prone to excessive mucus and suffering from a skin condition, which had left him partially bald. His partner Mathilda was in a similarly wretched state. "One can only hope that a century from now 'musk ox' will not be Chinese slang for a useless object that can't be disposed of," commented a *New York Times* editorial at the time.

Countless animals, from polar bears to platypuses, have been shuffled around the planet as political pawns, with varying degrees of success. In 1826, France received a giraffe from Egypt which put Paris "in the grip of an intense giraffe-driven craze." The animal's distinctive coat influenced high fashion, with women even wearing their hair à la girafe.

Still, pandas have remained the preeminent diplomat of the animal kingdom. "The giant panda's exclusive natural existence . . . and its visual identity make it a perfect tool to attract people around the world and to create—at least temporarily—a good image of China," wrote Falk Hartig, a scholar of diplomacy who has studied the subject extensively.

China's fluffy diplomats no longer come for free, however. Pandas are loaned to zoos at a cost of $1 million a year and tied into their captive breeding program. When I visited Edinburgh Zoo in 2014, the director of pandas, Iain Valentine, was hoping for the pitter-patter of tiny panda feet. Valentine's life was, as a result, dominated by panda urine. Pandas have cryptic pregnancies and the only way to gauge if Tian Tian was about to give birth was by monitoring her hormones in forensic fashion. She had been trained to urinate on command but was not always performing appropriately. Valentine confided that had he known how much time would be spent scrabbling around her pen trying to siphon up some of her prognostic pee, he would have designed their enclosure with it in mind.

In the end, Tian Tian disappointed everyone by reabsorbing her fetus. This is probably one of the panda's adaptations to ensure it only gives birth in favorable conditions. Had Sweetie delivered a baby panda, it would, however, have been the property of China and subject to its own $1 million rental fee. Any babies have to be returned to Chinese breeding centers after two years—sooner if the politics demand it. In 2010, two days after the United States decided to go forward with a meeting between President Barack Obama and the Dalai Lama, the Chinese government registered its disapproval by forcing the return of the two American-born cubs to China.

"It's all about politics and money," Kati Loeffler told me. "Panda breeding is a full-time, multimillion-dollar industry, particularly if one can convince the public that pandas are incapable of reproducing on their own."

It's not just pandas in foreign zoos that rake in the dollars. In China, panda tourism increased fivefold after the newly minted middle class embraced domestic tourism, and it's now a significant moneymaker for the city of Chengdu. At the breeding center, ardent fans pay $170 a pop for the chance to have their photo taken holding a baby panda; even more for the opportunity to muck out their pen.

Not all baby pandas have appreciated the adoration of their fans. In 2006, a tourist who paid to frolic with baby pandas suffered the painful embarrassment of her new fluffy friends suddenly turning on her. "She made the mistake of stroking them on the head a little too enthusiastically—and found herself suddenly knocked to the ground," the papers said. The woman was rescued, weeping and wailing, from the baby panda pen with only a damaged pride. A few months earlier, at the same center, another snap-happy tourist lost her thumb in pursuit of the perfect Kodachrome moment.

The panda's bamboo-munching lifestyle has equipped it with a fearsome bite, to crack its way into the tough sheath of the plant's stem. The powerful muscles in the panda's cheeks, which ironically give the bear its adorable big round head, deliver a bite force quotient that placed the panda in fifth place, wedged between the lion and jaguar, in a recent study of the top bite-forces among carnivores.

A website entitled When Pandas Attack has catalogued the surprisingly numerous victims of the panda's thirteen-hundred-newton bite. The wounded include a keeper at a Hong Kong amusement park who was mauled by a panda named Peace; a former French president who had to be rescued from the jaws of a male named Yen-Yen (narrowly avoiding a major diplomatic faux pas); and a drunk man who fell into a panda enclosure at the Beijing zoo and tried to give a repeat offender called Gu Gu a hug. The man woke up in the hospital having nearly lost his leg. "I always thought they were cute and just ate bamboo," he later told CNN. Which just goes to show how perilous the great modern panda myth can be.

Until recently these panda attacks had been confined to captive pandas. But in 2014, an elderly Chinese man required over fifty days in the hospital after his leg was ravaged by a wild panda rampaging through his village near the Baishuijiang National Nature Reserve. Who knows, with the pandas' homeland increasingly invaded by humans, maybe we'll start to see more of these wild attacks, which, seen through anthropomorphic eyes, could be the bear's bloody revenge for decades of misrepresentation, ridicule and rectal probes.

But my guess is this savage side of the panda is unlikely to ever tarnish its wholesome image. We like our pandas harmless and helpless. Such is the power of cute.

CHAPTER 12

PENGUIN

Order Sphenisciformes

*All the world loves a penguin: I think it is because
in many respects they are like ourselves, and
in some respects what we should like to be.*
—Apsley Cherry-Garrard, *The Worst Journey in the World*, 1910

Evolution has done a splendid job of equipping the penguin for a life chasing fish in the frigid sea, but bird biology requires they return to dry land to lay eggs, raise chicks and roost. To accommodate life on sea ice, where we generally picture them, penguins are built like a Thermos. But not all penguins spend their lives skidding around on the ice; half of the existing species inhabit far cozier climes as far north as the equator. For those penguins that inhabit more tropical locations, waddling around in a thick-feathered wetsuit can be a perilous sport.

These birds have evolved creative strategies to avoid getting overcooked. Some species stand around panting like dogs, others are forced to seek out shade. Yellow-eyed penguins trek over half a mile inland (quite a marathon for those little legs) to raise their chicks in the cool of the New Zealand rainforest. Galápagos penguins avoid the brutal equatorial sun by nesting in uncomfortable-looking cracks of coastal lava rock. Humboldt penguins arguably have it even worse. Living on the barren coast of Peru they are forced to improvise shade by carving castles of crap from piles of their own mature manure. Fairy penguins solved the problem by shunning the sun altogether and becoming nocturnal.

My first encounter with Antarctica's most famous resident was on a balmy golden sand beach, just a short drive from Melbourne. The southern coast of Australia is home to several colonies of fairy penguins. Standing at almost twelve inches tall, they are the planet's littlest penguins, with the biggest following. Tourists have been flocking to Phillip Island to watch the petite penguins since the 1920s. I joined several hundred ardent fans, some clutching freshly purchased plush penguins several times larger than the real thing, to wait for the arrival of the island's famous penguin parade, a nightly carnival in which the tiny, shiny blue birds totter up the beach from the surf to their sandy burrows as soon as the sun goes down.

The penguin parade was billed with great pride by the local tourist board as "a waddle on the wild side." The diminutive birds did not disappoint. As the warm Australian sun dipped below the horizon, the ocean surf began spluttering out dozens of pint-sized penguins. As they shuffled up the beach, it was impossible not to watch them and smile.

The penguins' farcical walk is quite deceptive. Those stiff feet, so ill at ease on land, act as a rudder underwater, allowing penguins to make hairpin turns at speeds of over thirty miles per hour. They are the fastest maneuverers and deepest divers of any bird; emperor penguins can reach depths of over sixteen hundred feet. These seabirds spend 80 percent of their lives as slick predators, but we only get to see the 20 percent spent staggering around on land like Charlie Chaplin.

"Our perception of animals is based on where we are able to observe them," Dr. Rory Wilson explained. He's the genius who has fitted hundreds of penguins with speed-o-meters, beak-o-meters and even bum-o-meters in an attempt to uncover their underwater life. "Seeing penguins stumble around being failures on land is like seeing the world's greatest athletes stumble around in the dark and never realizing what they are capable of," he told me. "It's impossible to swim like a penguin *and* run like a cheetah on land."

The muscles that control penguins' feet have to stay warm to function, so they're hidden under feathers way up in a penguin's leg. They

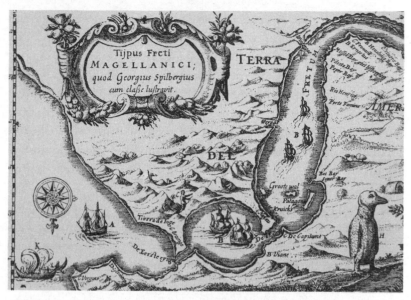

One of the earliest maps of the Strait of Magellan (sixteenth century) features a mellow-looking penguin out for a stroll—an indicator that these fat, flightless ready meals were up for grabs along those shores.

maneuver their extremities by a remote "pulley" system that's about as efficient as operating a Muppet and that gives the penguin its distinctive wobble. This quirky act of pathos has blinded us to the penguin's true story.

THE FIRST PENGUIN to be described by Europeans wasn't a penguin at all. It was a great auk. To be fair to the sixteenth-century captain who goofed up, the physical ID of a great auk does sail pretty close to that of a penguin. They are similarly fat, flightless, black-and-white birds that inhabit large rookeries on isolated rocky islands, albeit in the polar opposite, *northern* hemisphere. And they shared one other crucial characteristic: they were very easy to catch.

These portly birds became a boon for the starving sailor. Sir Francis Drake wrote about an island in the Strait of Magellan where he had killed three thousand "fowl which could not fly" and which were "the bigness of geese."

Mythical "Penguin islands" featured on maps like buried treasure, such was their significance to the survival of men at sea. From the time of Drake onwards, the word "penguin" became common currency for waddling ready meals, regardless of whether they were found on islands in the northern or southern hemisphere. According to connoisseurs, these "penguins" tasted like fish if cooked with their fat. If the fat was removed, their flesh (perhaps with a little wistful imagination) could pass as beef. They boasted the added novelty of their fatty carcasses being so flammable that their bodies could be used as their very own barbecue. Wrote one able seaman in 1794: "You take a kettle with you into which you put a Penguin or two, you kindle a fire under it, and this fire is absolutely made of the unfortunate Penguins themselves. Their bodies being oily soon produce a Flame; there is no wood on the island."

Despite their superficial form, and perhaps flavor, great auks are part of a completely different family of bird, more closely related to guillemots and puffins than penguins. Their similarities are, however, a great example of convergent evolution; namely, when two very different and unrelated animal groups evolve the same solution to a survival dilemma. In this case, both birds evolved to fly underwater to feed on small fish and sea creatures. They ditched the large, fragile wings and light-boned bodies that favor conventional airborne flight to become blubber bullets with short, powerful flightless flippers and squat, streamlined bodies—a shape so effective that nothing designed by humans has ever managed to beat the penguin's low drag coefficient. They also developed the same tuxedolike livery for camouflage: their white front disguises them from predators and prey looking up at the sun-bleached surface of the water, and their black back conceals them from predators above against the murky depths below. Add to this the same webbed feet and short legs that waddle so ineffectively on land. You can see how mistakes might arise, especially among sailors semidelirious with hunger.

Great auks were eventually christened, somewhat unhelpfully, *Pinguinus impennis*, meaning "featherless penguin"—of which they are neither. This inappropriate name did nothing to quell confusion between the two monochrome seabirds, which burbled on for centuries. This so annoyed Georges-Louis Leclerc, Comte de Buffon, that he proposed the penguin be renamed. The French aristocrat could have chosen "arse-feet," which was the moniker adopted by certain sailors after watching the birds swim along with their legs trailing behind them. But for reasons best known to himself, he decided on *manchot*, or "one-armed" in French. Since the penguin, like all birds, has two quite distinct wings, or arms, the name failed to catch on. Luckily for the Comte (if not for the birds), great auks were eventually eaten to extinction, bringing some form of closure to his personal penguin stew.

But confusion over what type of animal the penguin was continued. Some early explorers thought they were part bird, part fish. Others saw them as some kind of missing link between dinosaurs and birds. Having spent many hours marveling at their peculiar reptilian feet, which appear to have been stolen from a crocodile, I can understand why. But this turned out to be a surprisingly dangerous misconception.

The chief architect of the theory was the polar explorer Edward A. Wilson, who had served as ornithologist on Captain Robert Falcon Scott's famous expedition to Antarctica on the *Discovery* in 1901–4. Edward Wilson was a highly respected pioneer of penguin research whose observations helped crack the mystery of the emperor's rather unenviable reproductive cycle, in which the male weathers the brutal Antarctic winter with an egg on his feet and nothing to eat while the female, her reserves dangerously depleted from egg laying, spends two months gorging herself at sea. The male and female then take turns raising the chick and feeding themselves—a relay of extreme endurance that Wilson described as "eccentric to a degree rarely met with even in Ornithology." He believed the emperor was a relic species whose eggs held the secret to their evolution. In his report on the Antarctic mission, he declared: "the possibility that we have in the emperor penguin the nearest approach to a primitive form not only of penguin but of a bird makes the future working out of its embryology a matter of the greatest possible importance."

Wilson's ideas on embryology were influenced by the German biologist Ernst Haeckel. In 1868, Haeckel had devised the rather neat (if sadly not true) notion that all animal embryos go through stages of development that mirror their evolution from distant ancestors—that, as he would be famously paraphrased as saying, their ontogeny (individual development) recapitulates their phylogeny (species development). This grand "theory of recapitulation" was artfully illustrated by the zoologist himself, whose exquisite, if controversial, drawings of developing fetuses were a highly persuasive piece of promotion.

Assuming Haeckel was right, Wilson believed the emperor penguin's egg would be a time machine, transporting him back to lost transitional stages in the evolution of reptiles to birds. "The earliest bird, the *Archaeopteryx*, had teeth," Wilson explained in a lecture on penguins in 1911. "One hopes to find real teeth in the embryo of the emperor penguin, though none are present in the adult bird." Wilson also wanted to see if the papillae that develop into penguin feathers corresponded to those that become a reptile's scales. This was just fifty odd years after Darwin's shocking theory of natural selection was first published, and not everyone was converted. Wilson hoped that the emperor egg was just what was needed to crush the naysayers and prove Darwin's theory to be true.

He somehow managed to persuade Captain Scott that his far-fetched mission should be an essential part of the scientific endeavors of the explorer's second Antarctic expedition. So in June 1911 Wilson, Henry "Birdie" Bowers and Apsley "Cherry" Cherry-Garrard set off on a quixotic crusade to find lost dinosaur teeth in an emperor's egg at the ends of the earth. Their 124-mile side trip from base camp was later coined "the worst journey in the world" by Cherry-Garrard, its sole survivor. And he wasn't exaggerating. His unflinching memoir of the same title recounts the full horror of this ill-fated egg hunt.

Emperors nest in the middle of the Antarctic winter, so the three men were forced to make their expedition to the only known rookery, located at Cape Crozier, the easternmost point of Ross Island, in pitch darkness, navigating only by candlelight—no easy task when battered by Antarctic gales. The team took turns tumbling into the copious crevasses that littered their path. Temperatures dropped below −76 degrees Fahrenheit and made the snow so sticky they could pull only

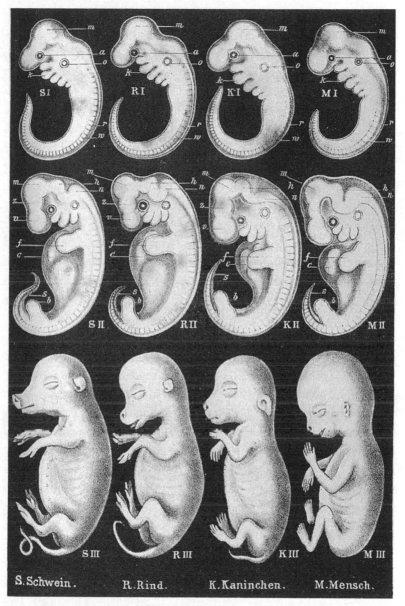

Ernst Haeckel's fine drawings of comparative embryos—in this case, a hog, calf, rabbit and man—provided persuasive PR for his flawed recapitulation theory (which sent a trio of explorers to the ends of the earth, with only one coming back).

one sledge at a time. They were forced into a trudging relay, gaining only one mile for every three they walked. Sweat turned their clothes into icy armor and their breath froze their balaclavas to their faces. Cherry's teeth chattered so violently they shattered, and the liquid in his blisters turned to Popsicles.

So horrific was the shlep, that by the time they reached their goal, Cherry had lost the will to go any farther: "I for one had come to that point of suffering at which I did not really care if only I could die without much pain. They talk of the heroism of dying—they know little—it would be easy to die . . . the trouble is to go on."

The frozen trio forced themselves forward, scrambling up nearly two-hundred-foot ice cliffs in complete darkness to reach the rookery.

The penguins weren't exactly pleased to see them. "The disturbed emperors made a tremendous row, trumpeting with their curious metallic voices," recalled Cherry. The explorers snatched five eggs from between the rowdy penguins' legs and skinned a few of the birds for their fatty fuel. But before they could say "mission accomplished," their fortunes took a turn for the worse.

The group became lost. Groping around in the dark trying to locate their trail, Cherry's frozen fingers dropped two of the eggs. By some fluke the explorers found their way back to their camp at the foot of (the aptly named) Mount Terror, and they immediately set about trying to get warm. It was the eve of Wilson's birthday and they fired up the penguin blubber stove, but, as if the birds were seeking their revenge, "it spouted a blob of boiling oil into Bill's eye." Wilson was blinded and he lay all night "unable to stifle his groans, obviously in very great pain."

"I always mistrusted that stove," said Cherry. But worse was still to come. A furious storm, "as though the world was having a fit of hysterics," blew their tent and most of their supplies away. The trio were forced into a makeshift shelter with a canvas roof "whipped to little tiny strips" by the force 11 gales. The trio spent Wilson's birthday "face to face with real death." With no food or fire they huddled in their sleeping bags singing hymns and dreaming of tinned peaches while occasionally thumping the birthday boy to check for signs of life.

After two days, there was a lull in the blizzard and by some miracle Bowers found their tent. "Our lives had been taken away and given back to us," wrote Cherry.

The three men staggered back into base camp on August 1, 1911. Their frozen clothes had to be cut from their bodies and their fingers were virtually dead. They looked as if they had aged thirty years in the five weeks since their little egg-collecting expedition had begun. Cherry never fully recovered psychologically from the ordeal, wrestling with PTSD for the remainder of his life. Wilson and Bowers somehow bounced back. This only brought them worse luck, since it enabled them to join Scott's doomed polar trek. Both men perished along with the rest of the party on the return journey from the South Pole, leaving Cherry the sole guardian of the three precious emperor eggs and the associated honors of evolutionary science.

Cherry took his responsibility as "Custodian of the Sacred eggs" extremely seriously. When he made it back to London, he delivered them personally to the Natural History Museum in South Kensington. He may have anticipated a hero's welcome. Instead, he was met by a junior jobsworth who showed absolutely no interest in the specimens, famously barking at him, "Who are you? What do you want? This ain't an egg-shop."

Cherry later wrote to the museum to complain about his reception. "I handed over the Cape Crozier embryos, which nearly cost three men their lives and cost one man his health, to your museum personally and . . . your representative never even said 'thanks.'"

Unbeknown to the explorer, the source of this curatorial insouciance was a rather untimely paradigm shift in evolutionary thinking. While Cherry and his colleagues risked their lives for science on the sea ice, Haeckel's recapitulation theory had itself capitulated. Science had marched selfishly on, rendering the emperor eggs redundant.

Cherry spent much of the rest of his life pushing to have the embryos studied, but twenty-one years would pass before any results were published—and they were not worth the wait. The zoologist James Cossar Ewart, of the University of Edinburgh, surmised from microscopic slides of the embryos that scales and feathers do not, as

Wilson had hoped, share a common origin. Then a final blow came in 1934, when the anatomist C. W. Parsons concluded cuttingly that the eggs, harvested rather too early in their development, didn't even "add to our understanding of penguin embryology."

HAVING LOST HIS TOENAILS, teeth and much of his sanity to the penguins, the polar explorer could be forgiven for harboring a certain amount of resentment towards them. Yet Cherry-Garrard remained resolutely enchanted. His writing exudes nothing but warmth and respect for these birds. "The Adélie penguin has a hard life, the Emperor penguin a horrible one," he noted at the end of his tumultuous memoir. "I challenge you to find a more jolly, happy, healthy lot of old gentlemen in the world. We must admire them: if only because they are so much nicer than ourselves!"

Such is the anthropomorphic power of the penguin. "They are extraordinarily like children," wrote Cherry-Garrard, "full of their own importance and late for dinner, in their black tail-coats and white shirtfronts—and rather portly withal."

He was not alone in seeing them this way. Such childlike analogies became the lingua franca for penguins soon after they were first spotted. The most earnest eighteenth-century scientists writing to the Royal Society waxed lyrical about how the seabird had "at first sight, the look of a child, waddling along with a bib and apron on." Even the seamen of the seventeenth century, capturing penguins for the pot, were charmed by how they "stand upright like little Children in white aprons."

As with the panda, the penguin's wobbly vulnerability seemingly mimics a toddler's first steps and triggers our innate desire to nurture. Add to this a long-suffering life at the edge of existence and a natural aptitude for clownish physical comedy and you have a recipe for anthropomorphic superstardom.

The potent mix of stoicism and slapstick made the first penguins to leave Antarctica an instant hit. The *Times*, reporting in 1865 on the penguins at the zoological gardens in London's Regent's Park, reveled in the birds' clumsiness, which was accompanied by a "ridiculous air of gravity." Their dignity in the face of endless icy pratfalls, combined with their unexpected love of tobogganing, made them stars of the moving image from the moment it was invented. The Charlie Chaplins of the animal kingdom were perfect fodder for children's books; their naturally iconic black-and-white suiting was (like the panda's) an ad agency's dream. Penguin logos adorned everything from books to biscuit tins.

The birds were even adopted by the American religious right as a paragon of Christian family values, thanks to the Oscar-winning 2005 documentary *March of the Penguins*. "This is the incredible true story of a family's journey to bring life into the new world. In the harshest place on earth, love finds a way," intoned the voice of Morgan Freeman over a series of shots of a perfect-looking emperor penguin couple caring for their impossibly fluffy chick. From that moment, the film played pretty fast and loose with the truth. The scripted narration interpreted the emperor's annual hormone-driven trek across the ice flows to breed as an epic love story. It wasn't true, but it made for a box-office smash. Christian fundamentalists flocked to the message, holding up the emperors' struggle as an allegory for spiritual striving and their behavior as an example for humans.

The conservative film critic Michael Medved declared *March of the Penguins* to be "the motion picture that most passionately affirms traditional norms like monogamy, sacrifice and child rearing." A group called the 153 House Churches Network set up a *March of the Penguins* Leadership Workshop for members to discuss the impact of the film on their lives. "Some of the circumstances [the penguins] experienced seemed to parallel those of Christians," declared the organizer. Churches block-booked cinemas for their members, and as of this writing the film remains the second most-viewed documentary in US history (sandwiched somewhat uncomfortably between *Fahrenheit 9/11*, Michael Moore's critical dissection of the Bush administration's "War on Terror," and *Justin Bieber: Never Say Never*).

Penguins may indeed be models of upright social behavior—in a literal, biomechanical sense. Choosing a flightless, fish-eating bird as a moral role model is, however, far from ideal. Not only do most penguins fail to uphold traditional Christian family values, but some of their sexual activities would challenge even the most liberal of communities.

Most penguins, for a start, are far from monogamous. The worst offenders are those "romantic" stars of the big screen, the emperors, with a whopping 85 percent switching partners from one year to the next. They do, however, have a reasonable excuse: since they carry their eggs around on their feet, there is no nesting site, and therefore no obvious rendezvous point to meet a mate at the start of each breeding season. They must instead try to locate their previous year's partner by shuffling around in a massive scrum and yelling. At a party of several thousand penguins, all dressed the same, and with a severely limited window within which to locate a mate, one can appreciate the difficulty in staying faithful.

Monogamy, when it does occur, can be a rainbow-colored affair. In his comprehensive compendium of animal homosexuality entitled *Biological Exuberance*, the Canadian biologist Bruce Bagemihl reported on lifelong same-sex partnerships among Humboldt penguins. Bagemihl's book celebrated a spectrum of natural sexual liberation in more than 450 species from gay gorillas to boto river dolphins with a penchant for casually boffing each other's blowholes. Many of his observations had been kept in the closet by zoologists for years, as they did not fit into neat Darwinian thinking. Incidents of mutual fellatio among male orangutans, for example, were explained away by one prudish biologist as "nutritively rather than sexually motivated." It is only very recently that the natural diversity of sexual behavior found in the animal kingdom has been accepted, with new theories about how such sexual activities diffuse tension, benefit child-rearing or simply might be *fun*.

Same-sex penguin partnerships have been particularly well documented in zoos, with some couples becoming famous for flying the rainbow flag. Dotty and Zee, two males at the Bremerhaven Zoo in Germany, recently celebrated their tenth anniversary together and have even adopted and raised a chick together. The penguins' celebration of sexual diversity may not sit well with buttoned-up Christian

A couple of gay Humboldt penguins at Bremerhaven Zoo in Germany were caught trying to incubate rocks as if they were eggs. The zoo flew in females from Sweden to "test" the penguins' sexual orientation—a move that enraged gay rights activists. The "Swedish strumpets" could not, however, break up the couple, who went on to foster their own chick.

conservatives, but it has also brought some disappointment to their more liberal champions. Roy and Silo, probably the penguin world's most famous gay couple, inspired a children's book, *And Tango Makes Three*, much loved by the LGBT community after they too raised a chick, in New York's Central Park Zoo. A few years later, Silo forsook his partner of six years and ran off with a female named Scrappy, which, according to the *New York Times*, "rocked the gay scene." Proving that the penguin's struggle with lifetime fidelity is very real, whatever the flavor.

Penguin divorce rates do tend to lower as you travel north from Antarctica, where warmer weather allows for a more flexible breeding season. This makes the penguin's mission to procreate less pressurized, so they can spend more time seeking out last year's successful partner. Galápagos penguins, which live as far north as the equator, manage to be the most faithful, with 93 percent of pairs reuniting each season. This fidelity may or may not help improve their ability to co-ordinate cooling efforts—and prevent the kids from turning to a crisp under the fierce equatorial sun—but you have to hope that it does.

Even those penguins that appear to stay together season after season may not be as faithful as they seem. Nearly a third of female Humboldt penguins cheat on their partners, often with members of the same sex. One in ten female Adélie penguins also have a bit on the side. Such female infidelity was generally thought to increase the genetic strength of offspring, until Dr. Lloyd Spencer Davis of the University of Otago, New Zealand, discovered that their motives may be a little more complicated than that. Adélie penguins, he claims, are one of the only animals on the planet to have turned to prostitution.

Adélies are your classic, knee-high cartoon penguin. They are the southernmost breeding of any bird and gather in vast rowdy crowds at the start of each brief summer to nest along the edges of the Antarctic Peninsula. Towards the end of the season, as the weather warms, there is a danger of the penguins' simple stone nests becoming flooded and drowning their eggs in meltwater, so the females go on the hunt for fresh pebbles to shore up their parental investment. Stealing is rife and scuffles are commonplace. "They can be surprisingly vicious, pecking and beating each other repeatedly with their flippers," Davis told me.

Some sneaky females have learned to avoid getting beaten up by possessive pebble owners by targeting the nests of unsuccessful males living at the edge of the colony. With no parental duties, these singletons are free to go on a pebble-hunting extravaganza and amass veritable stone castles. They are also extremely desperate to spread their seed. The sly female shuffles up to one of these lonely males with a deep bow and a coquettish sideways glance, as if she wants to copulate with him. The male bows back and steps aside to allow the female to lie down on his pebble castle and prepare to procreate. The sex is a swift affair, with the inexperienced male frequently misfiring and missing his target. After the deed is done, the female toddles back to her nest with a pilfered pebble in her beak.

Female Adélie penguins are one of the only species known to exchange goods for sex. Females dupe lone bachelors into having a quickie only to make off with some hard currency— stones—that they need to amass in order to protect their nests from drowning in frigid meltwater.

Davis noticed that some especially cunning females stole stones from the males without even offering sex in return. They flirted exactly as before, only they skipped the sex part and simply made off with a stone. "She takes the money and runs," as Davis put it. In response, the males were never seen to put up a fight—although some did make a desperate attempt to claim their conjugal rights as the female beat a hasty retreat with her booty. One particularly effective hustler was recorded swiping sixty-two stones within an hour.

Davis told me that the females have learned that the males "are not exactly dumb, as much as they are desperate." They have big stone nests and not much to lose. If there is a chance the female will have sex, then it is worth taking the risk. They may look foolish but, as Davis admitted, "Evolutionary-wise, it's a pretty smart move."

Cases of transactional sex are surprisingly rare in the animal kingdom—and hotly contested. The only other concrete examples Davis could find among vertebrates were chimpanzees (which have been observed bartering meat for sex) and, erm, ourselves. Which makes these female penguins a little more human than we may have expected, though perhaps not in a way the Christian right would be keen to discuss in Sunday school.

THE BEHAVIOR OF the male Adélie is, however, even worse. So shocking, in fact, that the Natural History Museum in London refused to share with the public the first scientific account of their sexual antics.

The private lives of the Adélies would have been lost to science were it not for a chance discovery in 2009 by Douglas Russell, the museum's senior curator of bird's eggs and nests. He was sifting through boxes of ancient documents, researching Scott's doomed second expedition when he happened upon a scientific paper from 1915 that caught his eye. Its prosaic title was "Sexual Habits of the Adélie Penguin" but "it had 'Not for Publication' written in large unfriendly letters at the top of it," he remembered. This, of course, "immediately piqued my interest," he said.

His interest was repaid, so to speak. The long-lost document was a wide-eyed account of male Adélies having sex with basically anything that moves. And quite a few things that don't. Like dead penguins. Not even fresh ones at that, but frozen relics from the previous mating season.

The lurid details of this avian orgy were delivered with deadpan Edwardian horror, the penguin perpetrators identified in decidedly human terms. They were "gangs of hooligan cocks" whose "passions seem to have passed beyond their control" and whose "constant acts of depravity" run the gamut of masturbation, recreational sex and homosexual behavior to gang rape, necrophilia and pedophilia. Chicks were "sexually misused by these hooligans," including one who "misused it before the very eyes of its parent." Stray chicks were crushed and "very often suffer indignity and death at the hands of these hooligan cocks."

The author, Dr. George Murray Levick, was a famous pioneer of penguin research. As a surgeon and zoologist on board Scott's second

expedition, he'd been afforded the rare privilege of observing the Adélie penguins at Cape Adare for twelve whole weeks during the Antarctic summer of 1911–12. He is to this day the only scientist to have spent
an entire breeding season studying the world's largest colony. His
meticulous daily observations were published by the Natural History
Museum upon his return in 1915 under the authoritative title "Natural
History of the Adélie Penguin," but the more unusual sexual procliv-
ities of the penguins were nowhere to be found among the pages of
Levick's definitive work.

Russell wondered why. After a fair amount of digging he unearthed
a note from the then keeper of zoology at the Natural History Museum to the curator of birds that instructed a conspiracy of silence
around the Adélie's sexual secret. Curiously, the note said, "We will
have this cut out and some copies printed for our own use."

Levick's graphic account of the sexual behavior of Adélie penguins was a little too ripe for the tastes of the post-Victorian academic
world. This was, after all, an era in which prudery dictated that verbal
or written communication about sex or emotions be couched in the
language of flowers, that the word "leg" was too naked to be used in
public and that any hint of homosexuality was positively wicked. Per-
verted penguins were not something polite society was quite prepared
for. The pamphlet was privately circulated, like some kind of under-
the-counter penguin porno, among a select group of peers who were
deemed learned and discreet enough to handle its graphic content.
Only one hundred were ever printed. "It's a miracle the copy I discovered survived," Russell told me.

Further detective work uncovered Levick's original field notebooks,
which exposed the full extent of the Adélies' behavior and the good
doctor's resulting turmoil. His initial observations of the first penguins arriving at the colony were laced with wonder. But after witness-
ing the penguin's "astonishing acts of depravity," he began to encode
his more lurid observations in ancient Greek—an old boarding-school
trick for keeping secrets.

One coded section described a male penguin "actually engaged
in sodomy" upon one of its own species. "The act occurred a full

minute," Levick noted conscientiously. But he made no attempt to explain this behavior. His scientific analyses were paralyzed by his horror. He solemnly concluded: "There seems to be no crime too low for these penguins."

Levick's observations of the Adélies were decades ahead of his time. It wasn't until the 1970s, sixty years later, that the penguins' dirty secret was discovered by another visiting scientist and made public. Then, the behaviors were recognized as a regular part of penguin life, triggered by the pressure of the short breeding season.

Adélies gather at their colonies in October. They are flooded by hormones and have only a few weeks to find a mate. The young males have no experience of how to behave, so many respond to inappropriate cues, exhibiting a somewhat flexible interpretation of what constitutes a potential mate. "It's not as if a young Adélie penguin is wandering through the colony and sees a frozen female and thinks, 'I've always quite fancied having sex with a dead frozen female,'" Russell said. "You have to be very careful about placing value judgments on animals. People are always keen to draw analogies with human behavior, but you've got to remember, it is just a bird with a very tiny brain."

To these hormonal penguin teenagers, a frozen penguin lying with its eyes half open is very similar in appearance to a compliant female.

"Anyone who knows birds knows they are quite famous for this sort of thing," Russell explained to me with an air of thinly veiled exasperation. So I checked out an online birdwatchers' forum to see what they had to say.

A thread devoted entirely to avian necrophilia had postings of all manner of shenanigans, including a "feral pigeon" mounting "a dead house martin," which, the witness added helpfully, is "a much smaller bird." Not to mention a different species. Another birdwatcher reported seeing a female house sparrow, a road victim unfortunately squished with its wings splayed out, "as if presenting to a male." This indeed turned out to be a deadly turn-on for the cock that flew down to mate with her and "was duly squashed in its turn." There was also a posting about an aggressive encounter between two cock pheasants, wherein one attacked another that had been hit by a car. "The episode ended with the first pheasant mounting the dying one and copulating with it," the post read. ("Grim stuff, but I have photos if anyone is interested!" the person added cheerily.)

The Adélies' urge to reproduce is so intense that when a researcher set out a dead penguin frozen in such a position, many males found the corpse "irresistible." It soon was so damaged by its repeated deployments, all that remained was "a frozen head, with self-adhesive white O's for eye rings." This was still "sufficient stimulus for males to copulate and deposit sperm on the rock"—a sight that would have probably tipped poor Levick over the edge.

There are still great limits in our understanding of animals. "From an evolutionary point of view, in a species in which there is a very limited opportunity for breeding, this behavior has purpose," Russell said, before adding with a laugh, "There's no romance here at all."

CHAPTER 13

CHIMPANZEE

Species *Pan troglodytes*

> *A brute, but a brute of a kind so singular, that man*
> *cannot behold it without contemplating himself.*
> —Comte de Buffon, *Histoire Naturelle*, 1830

I've had more than my fair share of extraordinary animal experiences, but one will stay with me forever. I was in the Budongo Forest of Uganda filming for the BBC. I had joined a research team that had been studying a particular troop of wild chimps for almost a decade, shadowing their lives from dawn to dusk every single day until they'd become habituated to the scientists' presence and now ignored them. For me, this promised a rare peek into the intimate lives of our nearest cousins.

But first we had to find them. That meant starting our expedition at the tail end of the night in the hope we could catch the chimps as they woke in their "sleep tree" and before they disappeared into the depths of the forest.

Sneaking into the sleeping jungle was an exercise in sensory deprivation. Under the cover of the canopy, it was dark, still and uncannily quiet, the steady *thwick-thwack* beat of rubber boots—standard issue to protect our ankles from the toxic retaliation of any snoozing snakes we might disturb—providing a spartan soundtrack for our thoughts. But sunrise is fast in the tropics, and before too long the first shafts of light began to illuminate the morning fog with a warm yellow glow, revealing the riot of life all around us.

I've always felt that the rainforest is my cathedral, the place where I feel closest to my god—evolution. And Budongo is an impressive place of worship: 310 square miles of dense jungle that hugs the eastern edge of the Albertine Rift, part of the Great Rift Valley where man himself is thought to have evolved. It is the largest indigenous rainforest still standing in East Africa, and although many of its magnificent mahogany trees were uprooted by the Victorians to deck out the Royal Albert Hall, a smattering of old-growth trees still stand, some twenty stories tall and almost half a millennium old.

Filing silently through the mists beneath these ancient trees felt like stepping back in time. Then came a crescendo of distant pant-hoots. This rising *whoop* that signals a chimp's excitement can penetrate the forest for miles and seems to go right through you. Hearing it gave me goose bumps—much the same effect it can have on another chimpanzee. We were getting close. I felt a flush of adrenaline. Chimps have a fearsome reputation. Although reports of their being ten times stronger than humans (with the ability to rip off a man's arm with ease) are overblown (they are merely twice as strong), I still felt apprehensive about arriving on foot to wake them without so much as a banana smoothie as a breakfast offering.

The research team stopped underneath a towering fruit tree and pointed upwards. At first I couldn't see anything; the chimps' black bodies blended into the infinity of the forest. But my eyes gradually adjusted, and as though I were looking at a magic-eye painting, the chimps emerged from the gloom: a dozen of them studiously munching away on their morning repast. I'd seen chimpanzees countless times before, acting up at the zoo or playing sidekick to truck drivers or Tarzan on TV, but this was totally different. They were somehow familiar and unfamiliar at the same time; like us, but not like us. The effect was mesmerizing and strangely emotional. It was such a poignant scene, a window perhaps into our distant past, made all the more meaningful by the increasing rarity of seeing these endangered creatures in the wild today.

My reverie was shattered by the sound of a fart. Wild chimpanzees, it turned out, suffer from profuse flatulence—loud, loose and unrepentant, the sound of an animal that subsists on unripe fruit and doesn't give a fig about being polite. The team told me the sound of distant trumpets is one of the best ways of locating lost chimps in the vast expanse of trees. I was unprepared for this particular dawn chorus, which was more like a scene from a Mel Brooks movie than anything I'd seen in an Attenborough documentary.

WE CAN'T HELP but look for a reflection of ourselves in the animal kingdom, but in the chimp we find a mirror image that is disorientating in its familiarity. This has promoted confusion and fear, and helped ensure that our closest relative is also one of the most tragically misunderstood animal cousins. Our obsession with the line that divides us—where it lies and what happens if it is crossed—has ushered in some of the most misguided moments in science.

"The constitution of the ape is hot and since he is rather similar to man, he always observes him in order to imitate his actions," wrote the visionary German nun Hildegard of Bingen in the eleventh century. "He also shares the habits of beasts, but both these aspects of his nature are deficient, so that his behavior is neither completely human nor completely animal; he is therefore unstable."

It is unlikely that Hildegard had ever clapped eyes on an unstable ape herself. To the earliest naturalists, apes were essentially mythical beasts, their descriptions cobbled from hearsay and muddled up with tales of pygmies, satyrs and weird savage men who covered their bottoms with their own long ears, all trapped in an uneasy hinterland between humans and beasts.

In his great encyclopedia, Pliny the Elder claimed that apes could play chess, while medieval bestiary writers highlighted the ape's grave fear of snails. Each acknowledged the animal's eerie ability to imitate, or "ape," man. This made them devilish, for if man had been created in God's own image, this scary, hairy imposter must be his nemesis. The illustrations that accompanied these early texts were suitably surreal. One particularly oft-copied portrait depicted a large, hairy lady standing proud while sporting an impressive mane, massive pendulous udders and a walking stick.

The earliest European eyewitness reports of wild apes were no less bizarre. One of the first came from Andrew Battel, an English privateer captured by the Portuguese in 1589 and incarcerated in Angola. Battel spent eighteen years yo-yoing between prison and trade expeditions for his captors in Africa. When he finally made it home to Leigh-on-Sea, the wily Essex lad had quite the story to tell. He turned his misfortune into a bestselling adventure that included a lengthy, if somewhat fanciful, description of what are widely believed to be gorillas and chimps. Battel described "two kinds of Monsters, which are common in these Woods, and very dangerous" and then gave a fuzzy account of hirsute, humanlike creatures that he called "pongos" and "engecos"; they built tree houses and beat elephants with clubs.

Battel was not alone in being befuddled by hairy humanoids. The Dutch traveler Willem Bosman claimed that apes in West Africa attacked people and could speak, but chose not to, so they would not be forced into working, "which they do not very well love." He considered them "terribly pernicious sort of brutes, which seem to have been only made for mischief." Others said the creatures were known to kidnap children, ravish women and keep humans as pets.

The first live chimp arrived on British soil to much fanfare. The *Speaker*, an English merchant ship, docked in London in 1738, bringing with it "an animal of remarkably and terribly hideous countenance . . . called by the name Chimpanzee." Unsure of how to greet this new arrival, the British did what they knew best and furnished the animal with a cup of tea. It was said to have sipped it daintily, like a human. The chimp's eating habits were, however, less suitable for the Georgian drawing room. "It seeks food from its own excrement," screamed the press. In addition to these coprophagous tendencies, the chimp was reported to seek human females for "illicit sexual intercourse"—a recurring insecurity among Victorian visitors to zoos in later decades.

It wasn't just the animal's behavior that confused. The first attempt at dissecting a chimpanzee, conducted by the English physician Edward Tyson, revealed unsettling similarities between ape and man, chipping

A confusion of early apes and ape men as featured in "Academic Delights," edited by the great taxonomist Carl Linnaeus (1763). It was this beastly classification of man, putting us so close to creatures like Lucifer (second on the left), that made the Comte de Buffon so irate. For once, I can see his point.

away at our God-given sense of superiority. Of the chimp brain, Tyson noted in 1699: "One would be apt to think, that since there is so great a disparity between the Soul of a Man, and a Brute, the Organ likewise in which 'tis placed should be very different too." But it wasn't. Instead it bore a "surprising" resemblance to human gray matter.

At the time, specimens available for study were few and far between. The different great ape species—chimps, gorillas and orangutans—became confused and conflated. This made them hard to categorize. The father of taxonomy, Carl Linnaeus, initially divided the apes into two: the more and the less anthropomorphic. The former category he called *Homo troglodytes*, which he designated as a second species of humans (the "cave-dwelling man"), while the latter he called *Simia satyrus*, an altogether separate set (a "satyr monkey").

The uncomfortable proximity of man to these excrement-eating, lascivious beasts did not please the aristocratic Comte de Buffon, who poured characteristic scorn on Linnaeus's attempts to classify them. Buffon fashioned his own, even more eccentric solution: he assumed that chimpanzees—or "jockos," as he called them—were in fact juvenile orangutans. The fact that his adult was a giant ginger beast, and his child small with black hair, did not bother the Comte, who pointed out: "In the human species have we not similar variety?" The Laplander and the Finn, he argued, shared equal disparity, despite living in the same climate (which may have been a sly dig at his Scandinavian taxonomic rival).

Buffon devoted endless pages of his encyclopedia to exhaustive descriptions of the striking similarities between apes and humans, right down to their "fleshy posteriors." But these physical analogies didn't offend him. Instead he interpreted the apes' likeness to humankind, yet total absence of any "power of thought and speech," to be the ultimate proof that man had been "animated by a superior principle," which forged his preeminence over these brutes. Any system that gave the two neighboring classifications was, in his mind, deeply humiliating. Linnaeus took his revenge on the Comte by naming a plant *Buffonia tenuifolia*—a jab at the Frenchman's *tenuous* grasp of taxonomy.

By the time Darwin published *On the Origin of Species* in 1859, the search for some characteristic that set us apart from our ape cousins had become a matter of scientific urgency. Chief amongst the protagonists in the quest was one of natural history's most notable villains, Sir Richard Owen. As Britain's most celebrated anatomist, Owen had clawed his way to the pinnacle of the scientific establishment, even tutoring the Queen's children in zoology. But he was a man of such a jealous nature and cutthroat ambition that he was openly booed by his contemporaries.

A deeply religious man, Owen was a vociferous opposer of Darwin's evolutionary ideas. He could not accept that man was merely a "transmuted" ape. So he made it his mission to seek out the physical source of mankind's uniqueness. His first stop was the brain, which produced a trio of likely suspects including, most crucially, a small fold of skin at the rear known as the hippocampus minor. Owen claimed this innocuous bulge only existed in man and must therefore be the seat of human reason, and the source of man's "destiny as the supreme master of this earth and of the lower creation." This discovery provided Owen with the confidence to elect humanity to its very own highfalutin class, which he christened, in no uncertain terms, *Archencephala*, or "ruling brain."

When Darwin heard about Owen's claims, he mused in a letter to a colleague: "What a Chimpanzee [would] say to this."

It was the brash working-class biologist Thomas Henry Huxley who took the dissecting knife to Owen's theory in public. Huxley, who called himself "Darwin's bulldog," firmly believed that science should be separate from religion, stating, "He was not ashamed to have a monkey as an ancestor; but he would be ashamed to be connected with a man who used great gifts to obscure the truth."

He began to systematically inspect primate brains himself. Quickly he recognized the magnitude of Owen's error—and saw a delicious opportunity to "nail . . . that mendacious humbug . . . like a kite to the barn door."

In a series of public showdowns and scientific papers, Huxley exposed Owen as a deceitful plagiarist who had not only fabricated his theory by copying drawings of chimpanzee brains from other anatomists (whom he conveniently neglected to credit), but who had also ignored their blatant descriptions of the chimp's hippocampus minor. Huxley's own methodical dissections revealed the startling similarity between chimp and human brains. He declared Owen's theory to have been built like "a Corinthian portico in cow dung."

As it turned out, Owen was right that the secret to our very different lives may well reside in subtle differences in the structure of chimp and human brains; he just hadn't found them. In the face of Huxley's relentless attacks, he was forced to concede that the hippocampus minor was indeed present in apes. His reputation never recovered.

RICHARD OWEN WAS by no means the only scientist to fall from grace over the chimp. Scientific fascination with the boundaries—and lack thereof—between ape and man would dredge up an extraordinary cast of reprobates.

Around the start of the twentieth century, a Russian scientist named Ilya Ivanovich Ivanov was making a name for himself as the creator of strangely named beasts—zonkeys, zubrons and zorses—that wouldn't be out of place in the less credible corners of the medieval bestiaries. These animals were hybrids, genetic as well as linguistic bastards of a zebra and a donkey, a bison and a cow, and a zebra and a horse. Ivanov's greatest aspiration, however, was to breed a *humanzee*—a hybrid between a human and a chimp.

He wasn't the first scientist to dream of becoming a real-life Dr. Moreau. Interest in breeding humanzees was very much in the air. When the German physiologist Hans Friedenthal experimented with mixing blood from a human and an ape in 1900, he had found their antibodies did not attack one another and postulated that interbreeding might be possible between the two. In the following two decades, Dutch zoologist Hermann Moens and German sexologist Hermann Rohleder (author of the enigmatically titled *Die Masturbation*) both tried to test this prediction by inseminating chimpanzee females with human sperm. Neither got beyond the planning stage.

Ivanov's success at manufacturing zonkeys and zorses confirmed him as a pioneer in the art of artificial insemination, and his unique expertise made him the right man in the right place at the right time. During the 1920s, the newly established Soviet Union was keen to undermine religious thinking and prove the superiority of their willfully technocratic society. The Soviet authorities believed a hybrid would provide "extraordinarily interesting evidence for a better understanding of the problem of the origin of man" and "a decisive blow to the religious teachings and . . . in our struggle for the liberation of working people from the power of the Church."

It wasn't just heretical Bolsheviks who backed Ivanov's work. In 1924, the Pasteur Institute in Paris wrote to the Russian scientist to share good news: it would be "possible and desirable" for him to conduct his experiment at their newly established chimpanzee colony in West Africa. The iconic research center had developed something of a taste for crazy Russian scientists. They were already supporting Serge Voronoff, who was breaking equally bizarre ground in chimpanzee science—he claimed to have discovered the fountain of youth by grafting thin slices of chimp's testicle onto aging men's scrotum. He had

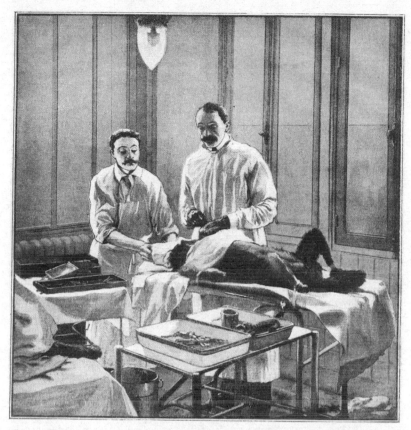

Voronoff's rejuvenation therapy made the front page of a French newspaper in October 1922, when the Russian pioneered his technique of grafting slices of chimpanzee testicle onto an old dog. Many more old dogs followed—of the billionaire variety—keen to add a little youth to their step.

come up with this rather imaginative "rejuvenation therapy" after observing eunuchs and undergoing some distinctly unenviable self-experimentation that included injecting his own testicles with a cocktail of crushed guinea pig and dog gonads.

Hand-stitching slivers of "monkey gland" using the finest silk thread was Voronoff's reassuringly expensive formula for extending human life up to 140 years. Although he insisted that "grafting is in no way an aphrodisiac remedy, but acts on the whole organism by stimulating its activity," the rumor was that it could restore a millionaire's flagging sex drive, as well as his memory and eyesight. In any case, it was enough to ensure that Voronoff's clinic was never empty. Hundreds of men signed up for the treatment—including Sigmund Freud who, having failed to find an eel's testicles, was clearly not afraid to experiment with his own.

Ivanov himself was in need of a little of Voronoff's monkey gland magic. The Pasteur Institute had offered him facilities but no funding, and, perilously low on cash, his humanzee project was wilting. So he stopped off in Paris en route to Africa to collaborate with Voronoff. They made headlines by transplanting a woman's ovary into a chimp called Nora and attempting to inseminate her with human sperm. No humanzees were conceived. Voronoff decided to stick with his lucrative day job of doctoring millionaires' testicles and Ivanoff flew off to French Guinea with only his son, a medical student, to support him.

On February 28, 1927, Ivanov attempted to inseminate a pair of chimps named Babette and Syvette with human sperm. Ivanov senior wanted to keep the experiment secret from his African assistants, and thus the situation was far from ideal: "The sperm was not completely fresh, but approximately 40 percent of the spermatozoa were moveable," he noted in his diaries. "The injection took place in a very nervous atmosphere and in uncomfortable conditions. The threat from the apes, the work on the open ground, and the necessity to conceal." The attempt was a failure.

Disappointment drove Ivanov into ever more desperate directions. He decided to change tack and lobbied the local governor to allow him to impregnate hospitalized women with chimp sperm. As if the ethical challenges thrown up by such an experiment weren't questionable enough, Ivanov proposed doing this without the women's knowledge, under the pretext of a medical examination. The authorities actually considered his plan, but eventually rejected it, a decision that, Ivanov noted in his diary, came as "a bolt from the blue"—which gives a sense for how divorced from reality he had become by then.

He was forced to move his mission back home, relying on his ability to acquire imported apes and consenting women in his homeland.

Indefatigable, he somehow managed to get both. But in the summer of 1930, the political winds changed and Ivanov fell victim to Stalin's widespread purge of scientists. He was arrested by the secret police, accused of counter-revolutionary activities and exiled to prison in what is now Kazakhstan. He died there two years later.

Could Ivanov's dream ever have been realized? I asked J. Michael Bedford, an emeritus professor of reproductive biology at the Weill Cornell Graduate School of Medical Sciences, who spent time in the 1970s investigating the early stages of fertilization, in particular how sperm attaches itself to the egg, with a view to developing a male contraceptive. He introduced human sperm to a range of different animal ova—from hamsters to squirrel monkeys to gibbons (a lesser ape). Bedford was surprised to find that human sperm has a high degree of specificity; the only egg it would attach to was that of the gibbon, our most distant ape relative. When I asked him what he thought would happen with a chimpanzee egg, he suspected a positive outcome: "Given that it is closer to man than the gibbon, it seems likely that sperm of the chimp would be able to fertilize a human egg and vice versa."

Fertilization is, however, just the first stage in a very long process fraught with potential failure. We may share 98.4 percent of our DNA with chimps, but producing a healthy humanzee baby would, according to Bedford, be "something of a crapshoot." He explained that in some hybrid combinations, live young are born sterile, but sometimes they are not; in others, the embryos begin to develop and then fail at some point in pregnancy. "It is not possible for me to predict whether the resulting embryo would survive," he said.

We may never learn whether humans and chimps could hybridize. But a study by researchers at Harvard Medical School and the Massachusetts Institute of Technology uncovered a ghostly secret in the human genome that suggests such liaisons may well have been part of our ancestors' past.

The scientists were comparing the genomes of humans and chimps using a "molecular clock" to estimate how long ago we diverged— the further back two species split, the more differences will have accumulated between their DNA sequences. The team estimated that humans and chimps separated no more than 6.3 million years ago, and probably less than 5.4 million years ago. But there was a massive anomaly with the X chromosome, which contained significantly fewer differences than the rest of the genome. The most logical explanation to the team was that speciation between humans and chimps had been "complex," which is a polite way of saying that the two emerging species continued to have sex and form hybrids for some time.

The amount of similarity on the X chromosome suggests that this wasn't just a solitary dark night at the disco but 1.2 million years of messy speciation. Nick Patterson, one of the team's lead researchers, told me how their discovery sent shock waves through the mainstream media. "The tabloids had fun. A headline in the German paper *Bild Zeitung* was '*Ur-Menschen hatte sex mit Affen*' (translation hardly needed) with a picture of the ugliest chimp they could find," he recalled. "But those who were shocked were missing the point. This is not a case of creatures like us mating with creatures very like chimps. We are talking about two groups of apes, one group very slightly more similar to us than to chimps."

While the idea of our ancestors getting down and dirty with the ancestors of our closest cousins may not appeal to some, the Harvard-MIT team think these hybrids could well have given our species an evolutionary boost, speeding up its adaptation to a new life out of the trees and on the savannah.

ANOTHER BOLD EXPERIMENT to blur the line between chimps and humans took place in 1960s America—but this time, the tinkering was behavioral rather than genetic. A newborn baby chimp was transplanted into a human family, to be raised as a human in isolation from its own species. The man behind this outlandish idea was Maurice K. Temerlin, a professor of psychology at the University of Oklahoma, who was particularly interested to see how the chimp, which he named Lucy, would develop socially and sexually.

There had been two previous attempts at raising chimps in human homes, but only as far as infancy, so to continue beyond puberty was uncharted territory. Temerlin probably never imagined he would wind up raising a gin-drinking chimpanzee teenager that took to masturbating with a hoover. But we know that's how his "experiment" panned out, thanks to Temerlin's earnest oversharing in his memoir, *Lucy: Growing Up Human*, which serves as an extraordinary time capsule of misguided 1960s pseudoscience.

"I am a psychotherapist. My daughter, Lucy, is a chimpanzee." So began Temerlin's self-indulgent account of his eleven years as a chimpanzee "father."

Lucy joined the family in a traumatic manner, when in 1965 Temerlin's wife, Jane, snatched the chimp, aged just two days, from her mother, a Californian circus entertainer. Temerlin considered this kidnapping the "symbolic equivalent of the act of giving birth." At the outset of their family "adventure," Temerlin wondered how human Lucy would become, and whether he, a self-confessed "Jewish momma's boy," could become a good "chimpanzee father." As the story unfolded, it became evident that the answer to his psychoanalytic hand-wringing was a deafening "NO."

Things started relatively innocuously. Lucy learned how to dress herself, use silverware and eat at the dinner table alongside the Temerlins' seven-year-old human son, Steve. Lucy was taught American Sign Language, eventually mastering more than one hundred words, including such essential chimpanzee terms as "lipstick" and "mirror." She even raised her own pet kitten. So far, so cute. But then comes a chapter entitled "Creative Masturbation," and things begin to get quite a bit darker.

At around the age of three, Lucy acquired a taste for alcohol after she swiped a drink from the nervous wife of a visiting academic. In the book, Temerlin *père* wrote about his guilt over giving alcohol to his teenage son, but curiously not about giving it to Lucy. Every night before dinner, he would "fix her a cocktail or two"—a G&T in the summer and a whiskey sour in the winter. Lucy eventually learned to work her way around the drinks cabinet and pour her own cocktails, which she liked to enjoy while lying on the sofa and flicking through magazines with her feet.

It was during one of these drinking sessions that Temerlin noticed Lucy getting creative with the nozzle of their Montgomery Ward vacuum cleaner. It was an inspired example of tool use, he noted—a skill once thought to be the exclusive preserve of man and a means of drawing that all-important line between them and us, until Jane Goodall observed chimps using sticks to fish for termites. Goodall's discovery led her mentor, Dr. Louis Leakey, to declare: "Now we must redefine tool, redefine Man, or accept chimpanzees as humans." One wonders if Leakey's redefinition would have been broad enough to include Lucy's imaginative hoover use.

The Freudian shrink in Temerlin was fascinated by his daughter's emerging sexuality, and whether Lucy was attracted to humans or chimps. Whereas many parents would have confiscated the vacuum cleaner and locked it in a cupboard, Temerlin dashed out to buy his daughter a copy of *Playgirl* instead of her usual favorite, *National Geographic*. Lucy did indeed take a shine to the magazine, fixating on the photos of naked men and stroking them so vigorously in key places that she ended up wearing through the pages. Pleased with his results, Temerlin next took the extraordinary step of dropping his trousers and joining his daughter in an afternoon of self-pleasure, "to see what would happen."

It is something of a relief to read that Lucy ignored Temerlin's looming onanism. In a chapter entitled "Oedipus-Schmedipus," the psychoanalyst dove into solemn conclusions about how his rejected tumescence was gratifying evidence of his status as Lucy's father and of her demonstration of an inherent incest taboo (one of his major research obsessions).

Eventually Lucy's behavior became too challenging for the Temerlins to cope with. She learned how to master all the locks in the house, often escaping into the neighborhood or locking herself in, and her parents out, of the house (perhaps not surprising, given all of the above). Temerlin claimed that his daughter even started to lie. When challenged for crapping on the carpet, Lucy's sign language response was to point the finger at Sue, one of his graduate student assistants.

Maurice Temerlin was sure to document "his daughter"
at work with her beloved hoover. His original caption for
the second picture reads: "Having achieved orgasm, Lucy
is enjoying a reflective moment before returning to her
magazine." Fortunately, there are no photos of Dad joining
in on the self-pleasure party.

At the age of twelve, a fully adult Lucy completely defied her parents' control. "Lucy was into everything," wrote Temerlin. "She could take a normal living room and turn it into pure chaos in less than five minutes." So with heavy hearts, the Temerlins realized their domestic experiment must come to an end and looked for a new family for their chimpanzee daughter.

And this is where Temerlin made his most catastrophic error of judgment: he decided to send Lucy home and set her free.

He packed Lucy off to a chimpanzee rehabilitation center in Gambia accompanied by a young scientist called Janis Carter, another of his graduate students. The African bush was a very long way from Lucy's suburban life in Oklahoma. Lucy had barely ever met another chimpanzee and had no desire to integrate into her new community. She didn't want to eat the wild leaves and fruit that the other chimps did, let alone sleep with them in the trees. She had acquired rather more sophisticated tastes and was now stranded in the jungle without a sofa or a cocktail cabinet. Janis Carter dedicated several years to trying to encourage Lucy to rediscover her chimpanzee roots in a valiant but ultimately futile effort. Eventually, the Temerlin's daughter was found dead, her hands, feet and skin removed. The suspicion was that she had been taken by poachers, whom she'd probably guilelessly approached, having no fear of humans. They took advantage of their unwitting and overeager prey. And that was Lucy's end.

FORTUNATELY, THE STUDY of chimpanzees has moved on from the narcissistic setups of the 1960s and now is primarily comprised of observations of our closest cousins in their natural habitat. This is significantly more challenging than watching chimps in a controlled situation, as I discovered when I joined Dr. Cat Hobaiter of the University of St Andrews and her team in the Budongo Forest of Uganda.

For a start, wild chimps can travel 6 to 12.5 miles a day in search of food. Keeping sight of them was like playing a game of hide-and-seek with Olympic-standard opponents. Cat was persistent. She believed that documenting chimpanzees' natural lives reveals not just more about them, but also provides a better template for studying the origins of human behavior.

"I think Lucy and other enculturated apes allowed us to ask what apes could do, under *extraordinary* circumstances. The answer is that extraordinary apes can do extraordinary things in extraordinary circumstances!" Cat explained to me. "Of course, there are very good ethical reasons why we would never want to replicate those studies today. But some of the same point holds in modern captive settings. You can test apes on puzzles, or in controlled circumstances that are not available in the wild. However, no matter how good the zoo or sanctuary, the environment these apes are in remains more humanlike than apelike."

Cat aims to study chimp behavior in its purest form, away from the polluting influence of humans. To accomplish this, she and any human tagalongs, like me, had to become invisible, figuratively. That means thinking like a chimp and following strict rules—first and foremost, no eye contact. Staring is an act of aggression among chimps and picking a fight with your study subject isn't a great move for those trying to be inconspicuous (or, for that matter, trying to avoid serious injury).

As we sat just three feet away from a family unit engaged in a group grooming session, the mother chimp looked up and caught me gawping at them. I followed Cat's guidance and immediately looked away. I picked up a leaf and pretended to inspect it intently while peering out of the corner of my eye to gauge whether my eyeballing had been noted. With great relief, I saw the mother chimp was still engrossed in picking and eating the ticks off her teenage sons.

Silence was the second requirement of Cat's fieldwork—a state I struggle to sustain at the best of times. What I hadn't anticipated was how hard it would be to hush my body language. Most intentional chimp communication is made up of subtle hand gestures and facial expressions. Their chatter is unusually peaceful, and I was struck by how they resided in relative quiet (aside from the farting). It is these gestures that Cat was trying to decode in order to compile the world's first chimp dictionary. Captive chimps like Lucy may have made the cover of magazines by mastering up to 250 words in American Sign Language, but wild chimps make do with far fewer. So far Cat had translated around seventy gestures for her groundbreaking glossary.

Many of these gestures are strikingly similar to our own. A handshake is a sign of affiliation, just as it is for any businessman closing a deal. I saw chimps use an upturned hand to beg for forgiveness and kiss to say hello. But it's dangerous to assume that our hairy black cousins are communicating just like humans. Cat constantly worked to shed her human preconceptions and think like a chimp. "We tend to think that chimps are very much like us. It's very easy to fall into that trap when analyzing the gestures—to assume, for example, that a handshake has a different meaning to an arm shake, because that makes sense to us," she said. "Maybe the chimps don't care what limb you shake. It all has the same meaning."

Some of the chimps' body language registers as quite the opposite of what it looks like to a human observer. "A bare-tooth grin means I'm nervous, worried or afraid," Cat told me. "That's the horrible thing about all those greetings cards where you have smiling chimps—they are definitely not *smiling*."

One of the most significant recent discoveries is that chimps tailor their communication according to what they think another chimp might already know. This insight into the minds of others—known as having a *theory of mind*—is a hot topic in animal psychology. It was long considered a unique attribute of being human, one of those key boundaries separating us from them. There have been numerous experiments probing for its existence in captive chimps. Cat showed me how colleagues of hers demonstrated it in the wild—an exhausting exercise that tested our own ability to read the minds of chimps.

The idea itself was relatively straightforward, if a little eccentric. We had to conceal a rubber viper along the path of a group of traveling chimps and, when the snake was revealed, observe whether the chimpanzee who saw it first communicated differently according to whether it thought the other chimps had seen the snake or not. Simple and elegant.

In practice, the experiment was anything but. We had to predict, in a vast forest, where the chimps would next go on foot. We then had to scramble through the thick jungle to get ahead of the apes in order to place the fake snake in their path (without them seeing). The snake was covered with a camouflage cloth attached to fishing line, at the end of which was a field assistant (also hidden) who was primed to pull it once the chimps ambled up. We had to hope that one of the chimps saw the toy-shop snake and believed it to be real—and then figure out whether their subsequent behavior was a warning to the rest of the group. Finally, we had to make sure we were in the right position to record the chimps' responses on a video camera.

We ran around the jungle for a whole day, getting ourselves utterly lost and savaged by ants as we crashed through the undergrowth. Only in the dying light did we finally manage to get our stars aligned. When the chimp that saw the snake (the last in the group) made an almost inaudible *hoo*—a soft warning—it suggested he "knew" the other chimps ahead of him had already seen the snake and he didn't need to make the loud barking alarm normal for such a threat.

The researchers did this experiment 111 *times* in order to draw their conclusions, which took them an exhausting six months.

The ability to read another's mind is extremely useful for chimpanzees whose lives are enmeshed in hierarchical but highly fluid social networks of up to one hundred individuals. Understanding the dynamics of this jungle soap opera is essential to survival, which means learning to recognize a lot of faces—or rather, in the case of chimps, bums. In a recent study by Frans de Waal, chimps were shown photos of their friends' rear ends and faces and were equally familiar with both. This makes sense for an animal that spends much of its life in a tree. "Most of the time you are looking up at bottoms," Cat told me. She has even developed genital photo flash cards to enable her research team to familiarize themselves with both ends of their study subjects.

After almost a decade together, Cat said she had come to know her chimps as well as her own family. "I have watched individuals growing up and so I have all that information available to me." By way of example, she told me about two brother chimps, Frank and Fred.

"I know that Frank, who is currently our up-and-coming alpha male, was a really outgoing young boy who would constantly push his luck with the adults when he was still young enough to get away with it. Compare with his brother who had the same mother and same rearing environment, same community, same forest but was a totally different chimpanzee. Fred was just incredibly quiet, laid back and not boisterous at all," she said. "Starting to unpick what the reasons are for that, how you get such radically different survival strategies, is fascinating."

Cat believes that the chimp's complex social life, intelligence and long lifespan are a recipe for individuality, a quality that she said is crucial to understanding chimps and indeed other species. "In Western science, we have a tendency to say we need to find a population of animals and we will take the average of their behavior and get rid of all the variabilities. Variation is bad and seen as a negative thing. Whereas individual differences are fundamental to looking at human behavior," she pointed out. "In our quest to get rid of variation we have totally overlooked the fact that that's a very interesting thing about animal behavior."

When Cat has visited chimps at other research sites across Africa, she's observed major differences among her chimps and these groups, "just as you'd find in people from India and Scotland," she said. "Females in West Africa, for example, are high ranking and integrated into politics in a way that they aren't in Budongo. When a female arrives, she is greeted as you would greet a male—a significant cultural difference."

Cat explained that talking about cultural differences among chimp populations in this way is relatively new. "There were a lot of people who treated chimp behavior in general as if it were the same. Now we understand that we were quite biased in what we thought of as "chimp" behavior, because for years much of [the data] came from Gombe"—where Jane Goodall did her pioneering research. "It turns out that those things are true for Gombe behavior, but they're not true for chimps in general. Even *within* Gombe—the group today and the group a couple of generations ago are very different, so we can see the impact of individuals, of personalities, and life history on group culture."

Many of these regional differences relate to tool use. A population of chimps from Senegal were recently observed living in caves and fashioning spears out of sticks that they sharpened with their teeth and then used to hunt bush babies hiding out in hollow trees. In Guinea, chimps were seen making leaf sponges to enable them to drink high-proof alcohol fermented from a palm plant. And in Uganda juvenile female chimps were documented seemingly playing with sticks as if they were toy dolls—cradling them and even making nests for them to sleep in at night. Each of these groups had developed unique tools, which were remarkably anthropoid in nature.

Perhaps the prize for the most curious case of tool use ought to go to chimpanzees in West Africa that were recently recorded hoarding rocks in neat piles—reminiscent of those uncovered by archaeologists at human sacred sites—and then hurling them at trees in an excited ritualistic display. Within a matter of days of the scientific paper on this extraordinary behavior being published, tabloids around the world were shouting about how chimps were "building shrines" at a "sacred tree." Could this be "proof that chimps believe in God?" asked one headline.

Dr. Laura Kehoe was one of the "baffled scientists" at the center of the media storm. "It was a bit ridiculous," she told me. "I think it's actually a really good example of how a myth can evolve and get completely out of control in a matter of days, to the point where you have religious people writing to you, thanking you for your work," she said. "I got this one letter—it was just amazing—from this woman in Ireland, to say how happy she was that chimps had religion and that she would be praying for me."

The original scientific paper had merely made the comparison with ancient human stone cairns. It had then pointed out that archaeologists should perhaps exercise caution when assuming that these human sites are sacred given their similarity to the chimps' handiwork. The chimp cairns, the authors proposed, might be associated with male chimp display or communication. The same chimps were known to use tree buttresses as drums to send long-distance messages, and their rock-throwing could be serving a similar purpose. The rock piles could also indicate something more symbolic, such as the marking out of territory. The authors noted the striking resemblance to indigenous West African stone shrines built at "sacred" trees in the same area, and how it would be interesting to explore the parallels in these behaviors.

As one of the paper's authors, Kehoe was approached by a number of online news outlets to write about the discovery and heavily encouraged to ponder big "what ifs?" around the idea that the chimp sites had spiritual meaning. She wrote a piece, and "the editor then changed the title from 'An interesting behavior found in our closest living relatives' to 'Mysterious chimp behavior may be evidence of sacred rituals,'" she told me. "Obviously once someone clicks on that title, that's what they want to read about and that's their whole mindset. It just got out of control from there."

It's a cautionary tale for scientists encouraged to sex up their findings in order to raise awareness for their work and for the urgent conservation needs of their dwindling study subjects. Thanks to the Internet, we have entered a new era of myth-making, where clickbait headlines and fake news are too often indistinguishable from the real thing.

While Kehoe didn't endorse the press reports that her chimps had found God, she did think they could have the capacity to experience feelings of awe. Other primatologists have suggested the same thing. Jane Goodall has said she's observed chimps behaving in an idiosyncratic, ritualized manner around waterfalls—throwing stones with their hair erect and then sitting down and staring at the tumbling water. Chimps cannot swim, so water is a danger, but this display is definitely not a fear response. It's something Goodall said might be "perhaps triggered by feelings of awe and wonder."

"The chimpanzee brain is similar to ours. They have emotions that are clearly similar to those that we call happiness and sadness and fear and despair and so forth," Goodall said. "So why wouldn't they also have feelings of some kind of spirituality? Which is, really, being amazed at things outside yourself."

Cat had seen a similar phenomenon in the Budongo chimps when they dance in the rain. "It's the most beautiful thing to watch. It only happens when you get a really big storm—that thundering rain that just drowns out every other noise and the chimps will do this weird ballet—like a display in slow motion, underwater, that's totally silent. It is totally different to any other display and they only give it in response to this very grandiose, big natural event. You know when you hear inspiring music and you just want to move. It doesn't feel like that's anything like religion, but it seems to be about awe for natural wonder. So perhaps there's capacity for it. I don't know."

As with so many animal mysteries, we may never know the truth. It could be the seeds of spirituality, or yet another instance of us viewing the animal kingdom through the prism of our own existence. I can't help but think that, given the alarming decline in chimp numbers at the hands of man, feeling connected with our closest cousins, rather than drawing more lines around what we think makes us superior, is the way forward in our relationship.

With each new discovery, the boundaries we have created around our uniqueness, century after century, continue to fade. The seventeenth-century bestiary author Edward Topsell wrote, in his definition of the ape: "They are not men, because they have no perfect use of Reason, no modesty, no honesty, nor justice of government, and although they speak, yet is their language imperfect; and above all they cannot be men, because they have no Religion, which (*Plato* saith truly) is proper to every man."

How much of that list remains true today of man or chimp?

CONCLUSION

We have much to learn from our centuries of misunderstandings about animals. Historians of science like to celebrate our successes, but I think it's equally important to examine our failures—especially when we ponder why the truth can be so completely unexpected to us.

The urge to anthropomorphize trips us up and obscures the truth. We are an insecure species, seeking reassurance about our behavior from drunk moose and busy beavers, and quick to condemn those creatures that don't conform to our moral code, like indolent sloths, cruel hyenas and dirty vultures. Our discomfort with the truth about these animals reveals a great deal about our hopes as well as our fears.

The constrictive moral codes of the ancient philosophers and the medieval bestiary authors are still perpetuated today in the popular press and natural history programs (some of mine included) that intone praiseworthy emphasis on traditional norms: heterosexuality, monogamy and the nuclear family—few of which really exist in nature.

That's not to say that some animals don't have a rudimentary moral compass. This is a hot topic right now, with researchers like the esteemed primatologist Dr. Frans de Waal pointing out that the basis of morality—a combination of empathy and a sense of fairness—is found in species as diverse as monkeys and rats. This suggests that moralizing could be part of our fundamental biological makeup.

But painting the animal kingdom with our artificial ethical brush denies us the astonishing diversity of life, in all of its blood-drinking, sibling-eating, corpse-shagging glory. We need not be afraid of these behaviors—they are not here to instruct us. Whether a penguin is gay, straight or having sex with a frozen head has no bearing on our own sexuality. Despite what we may think, we are not the center of the animal universe.

Which brings me to my second observation: if anthropomorphism is enemy number one, arrogance is a close runner-up. From obliterating beavers for their bogus medicinal balls to using frogs to test for pregnancy or recruiting pandas as diplomats, we have a history of viewing the rest of the animal kingdom as simply here to service our needs. This selfish standpoint has resulted in many of our most misguided mistakes. In these times of mass extinction, we cannot afford to make many more.

The quest for truth is a long and winding road, littered with deep potholes. Thankfully our methods are less brutal than those of our eye-popping past, but we are still stumbling along in the dark and making mistakes. With the rise of efforts to discredit science, there has never been a greater need for truth. Yet, wrong turns are an essential part of all scientific progress, which demands blue-sky thinking as it seeks out each new horizon in understanding. As long as our egos or dogmatic beliefs are not to blame, we should not be afraid to continue to make wondrous mistakes, like Charles Morton and his birds that migrate to the moon.

Acknowledgments

First and foremost, I want to thank Will Francis for being the best literary agent in the world and holding my hand so steadily through every stage of this book. The whole team at Janklow & Nesbit have been superb—PJ, Rebecca Folland, Kirsty Gordon and the rest. Thanks to them, I have been lucky to work with two expert editors—Susanna Wadeson and Thomas Kelleher—whose experience, ambition and endless patience have made this book something I am immensely proud of. Their respective teams at Transworld and Basic Books could not have been more supportive and I feel privileged to have them on my side. Special mention has to go to Robin Dennis, Elizabeth Stein, Christine Marra and Kate Samano for kicking the book up a notch with their meticulous copyedits, and Caroline Hotblack and Jo Ormiston for all the gorgeous images. And to Sophie Laurimore for keeping those TV plates spinning when I was buried in books.

The research for this book was immense. Matt Brierley and Joseph Russell did a great job helping me with initial forays into various animals. The Reaktion series of animal books were also a valuable first port of call. Thanks to the medical historian Jesse Olszynko-Gryn for being so generous with his research on amphibian pregnancy tests. And Henry Nicholls and Jon Mooallem for being not just inspirational writers but openhanded with their address books. But my greatest debt is to the scientists and conservationists who have shared their knowledge with me over the years. There are too many to mention, but you've met a few in the pages of this book. In particular I'd like to thank Dr. Andrew Crawford, Dr. Rory Wilson and Sam Trull (of the Sloth Institute) for casting an expert eye over various chapters. As did Dr. Mrinalini Erkenswick Watsa, who, along with Marianne Brooker, helped wrestle my chaotic notes and bibliography to the ground.

The writing of this book spanned two of the most challenging years of my life (mired by cancer and death). I could not have got through it without the help of my closest friends: Bini Adams, Heather Leach, Maxx Ginnane, Wendie Ottewill, Maxyne Franklin, Lisa Gunning, Chris and Tori Martin, Sara Chamberlain, Charlotte Moore, James Purnell, Luke Gottelier and Lesley Katon. Special mention has to go to Jess Search for laughing at my stories for so many years and encouraging me to push my creative boundaries. To Rebecca and Damien Timmer for inspiring me to become the Amphibian Avenger and subsidizing my metamorphosis with a home in their attic (and thank you to Jed and Samson for graciously sharing it). And, of course, a massive thank-you to Bruce for being Bruce, and propping me up with such thoughtfulness.

For their indispensable advice on how the hell to write a book, I am grateful to Alex and Natalie Bellos, Alexis Kirschbaum and Andrea Henry (who pushed me in the right direction at a key moment). Thanks to Jet for making the trial of having my photo taken so much fun; Archie Powell and James Brown for providing me with a place to live and work during a very tough summer; Tom Hodgkinson for letting me get up on the Idler stage to rave about sloths. And Liz Bisoux for all things web.

Finally, thanks to my mother, for allowing me to be whoever I wanted, supporting my crazy dreams and giving me my sense of humor (it's her fault there are so many testicle jokes in the book). And to my wonderful dad, for sinking that bathtub in the garden and stimulating my love of nature. Mum and I miss you terribly, but your warmth, wit and wisdom are woven through every page of this book.

Image Credits

127   Bat and incendiary device. (United States Army Air Forces.)
131   *Telmatobius culeus*, Lake Titicaca, Bolivia. (Pete Oxford/Nature Picture Library/Getty Images.)
137   Spermatozoon. Homunculus. Woodcut from "Essay de dioptrique" by Nicolaas Hartsoeker, Paris, 1694. (Wellcome Library, London.)
138   Mating frogs by Hélène Dumoustier. Ms. 972, BCMHN. (© MNHN (Paris)—Direction des collections—Biblothèque centrale.)
141   Audrey Peattie working at the family-planning laboratory in Watford hospital. (Reproduced by kind permission of Jesse Olszynko-Gryn, a medical historian based at the University of Cambridge and funded by the Wellcome Trust.)
150   Pfeilstorch. Lithograph by Friedrich Lenthe, 1822, Universitätsbibliothek, Rostock MK-865.55a. (Universitätsbibliothek, Rostock, Germany.)
153   "The Goose tree, Barnacle tree, or the tree bearing geese." Illustration from *The herbal or, generall historie of plantes* by John Gerard, London, 1633. (Wellcome Library, London.)
157   Fishing for swallows (*De Hirundinibus ab aquis extractis*). Illustration from *Historia de gentibus septentrionalibus* by Olaus Magnus, 1555.
173   A hippopotamus, "the inventor of Phlebotomy." Illustration from *Il Ministro del medico trattato breve*, part 2 of *Il Chirurgo* by Tarduccio Salvi, Rome, 1642. (Wellcome Library, London.)
175   Author with hippo. (Author's collection.)
179   Hippopotamus. Engraving from *The Gentleman's Magazine*, published in December 1772. (Private Collection/Photo © Ken Welsh/Bridgeman Images.)
188   "'The Elk falling down in an Epileptick fit being pursu'd by the Huntsmen"—from *A compleat history of druggs . . . The second edition* by Pierre Pomet, London, 1737. (© British Library Board. All Rights Reserved/Bridgeman Images.)
192   "An Antelope." Pen-and-ink drawing tinted with body color and translucent washes on parchment from "The Northumberland Bestiary," Ms. 100, folio 9, England, *c.*1250–1260. The J. Paul Getty Museum, Los Angeles. (Digital image courtesy of the Getty's Open Content Program.)
194   Intoxicated moose, Gothenburg, Sweden, 6 September 2011. (JANWRIDEN/GT/SCAN PIX/TT News Agency/Press Association Images.)
199   "Male Moose." Broadside, printed by G. Forman, New York, 1778. (Courtesy, American Antiquarian Society.)

206 Bear licking her cub into shape. Illustration from French bestiary,
     c.1450, The Hague, Museum Meermanno, 10 B 25, folio 11v. (Museum
     Meermanno, The Hague, The Netherlands.)

210 London Zoo keeper Alan Kent feeding Chi-Chi the giant panda, 29
     September 1959. (William Vanderson/Stringer/Hulton Archive/Getty
     Images.)

215 Staff of Chengdu Panda Breeding Base with twenty-three panda cubs
     born in 2016. Photograph dated 20 January 2017. (Barcroft Media/
     Getty Images.)

225 "One of the Earliest Maps of the Strait of Magellan." Engraving after
     sixteenth-century Portuguese map, from *The Romance of the River Plate*,
     vol. 1, by W. H. Koebel, 1914. (Private Collection/Bridgeman Images.)

229 Comparative embryos of hog, calf, rabbit and man. Lithograph after
     Haeckel from *Anthropogenie, oder, Entwickelungsgeschichte des menschen*
     . . . by Ernst Haeckel. Published by Wilhelm Englemann, Leipzig, 1874.
     (Wellcome Library, London.)

235 Gay penguins at the Zoo in Bremerhaven, Germany, February 2006.
     (Ingo Wagner, Epa/REX/Shutterstock.)

237 Adélie penguin with stone for nesting material. (FLPA/REX/
     Shutterstock.)

247 "Anthropomorpha." Engraving from *Amoenitates academicae, seu
     dissertationes variae physicae, medicae, botanicae* by Carl Von Linnaeus.
     Published by L. Salvius, Stockholm, 1763. (Wellcome Library, London.)

251 Serge Abrahamovitch Voronoff and his assistant operating on an old
     dog, according to his method of rejuvenation by grafting. Front page
     of *Le Petit Journal Illustré*, 22 October 1922. (Photo by Leemage/UIG
     via Getty Images.)

257 Lucy the chimpanzee with hoover. (Photographs taken from *Lucy:
     Growing Up Human* by Maurice K. Temerlin, reproduced courtesy of
     Science & Behavior Books, Inc.)

Notes

My reading list for this book was too long to reproduce in full. So apologies if you are hunting for the source of a specific fact; there were too many to cite every one. Instead I have given references for all direct quotations and a bibliography of key books and academic papers that informed the book. Interviews with experts were conducted in the field or while I was writing the book.

INTRODUCTION

2   **"not take it up alone"** Simon Wilkin (ed.), *The Works of Sir Thomas Browne, Including His Unpublished Correspondence and a Memoir*, vol. 1 (London: Henry G. Bohn, 1846), p. 326.

3   **"consists of a church-door key"** Sebastien Muenster, *Curious Creatures in Zoology* (London: J. C. Nimmo, 1890), p. 197.

CHAPTER I. EEL

7   **"animal concerning whose origin"** Leopold Jacoby quoted in G. Brown Goode, "The Eel Question," *Transactions of the American Fisheries Society*, vol. 10 (New York: Johnson Reprint Corp., 1881), p. 88.

7   **"proceeds neither from pair"** D'Arcy Wentworth Thompson (trans.), "Historia Animalium," *The Works of Aristotle* (Oxford: Clarendon, 1910), p. 288.

9   **"daily scenes of cannibalism"** quoted in Tom Fort, *The Book of Eels* (London: HarperCollins, 2002), p. 161.

9   **"comes out of the water in the night time"** Albert Magnus, *De Animalibus*, quoted in M. C. Marsh, "Eels and the Eel Questions," *Popular Science Monthly* 61.25 (September 1902), p. 432.

10 **"a smacking sound"** Bengt Fredrik Fries, Carl Ulrik Ekström, and Carl Jacob Sundevall, *A History of Scandinavian Fishes*, vol. 2 (London: Sampson Low, Marston, 1892), p. 1029.

10 **"thirty feet long"** Tom Fort, *Book of Eels*, p. 164.

10 **"yard and three quarters long"** Izaak Walton and Charles Cotton, *The Compleat Angler: Or the Contemplative Man's Recreation*, ed. by John Major (London: D. Bogue, 1844), p. 179.

10 **"the coffee-houses"** ibid., p. 194.

10 **More measured measurements** Fort, *Book of Eels*, pp. 166–7.

12 **"which he kept long"** Walton and Cotton, *Compleat Angler*, p. 189.

12 **"scrapings come to life"; "only way they breed"** Pliny the Elder, *Naturalis Historia*, book 3, trans. by H. Rackham (London: Heinemann, 1940), p. 273.

12 **"electrical disturbances"** Marsh, "Eels and the Eel Questions," p. 427.

12 **"reverend bishop"** ibid.

13 **"damnable an untruth"** Thomas Fuller, *The History of the Worthies of England* (London: Rivington, 1811), p. 152.

13 **"the progenitor of the silver eel"** David Cairncross, *The Origin of the Silver Eel: With Remarks on Bait and Fly Fishing* (London: G. Shield, 1862), p. 2.

14 **"I could not be expected to be acquainted with the names"; "employ names and terms of my own"** ibid., p. 6.

14 **"hair eels"** ibid., pp. 14–15.

14 **"tails of the horses"** ibid., p. 14.

14 **"looks about a bit"; "very troubled state"** ibid., p. 17.

14 **"act of parturition"** ibid., p. 32.

15 **"may seem strange"; "vegetable kingdom"; "Great Creating Gardener"** ibid.p. 5.

16 **"They believed me"; "rejoiced in the solution"** ibid., p. 27.

16 **"more palatable"; "dry mint, rue berries, hard yolks"** Richard Schweid, *Eel* (London: Reaktion, 2009) p. 77.

17 **"extreme importance"** Goode, "Eel Question," p. 91.

17 **"cry and beg"** Marsh, "Eels and the Eel Questions," p. 430.

18 **"eels keep no diaries"** Sigmund Freud to Eduard Silberstein, 5 April 1876, *The Letters of Sigmund Freud to Eduard Silberstein, 1871–1881*, ed. by Walter Boehlich, trans. by Arnold J. Pomerans (Cambridge, MA: Harvard University Press, 1990), p. 149.

18 **"tormenting myself"** ibid.

20 **"thin-head short-nosed"** Fort, *Book of Eels*, p. 85.

21 **"pathologically ambitious"** Fort, *Book of Eels*, p. 129.

22 **"A chance of luck"** Bo Poulsen, *Global Marine Science and Carlsberg: The Golden Connections of Johannes Schmidt (1877–1933)* (Leiden: Brill, 2016), p. 58.

22 **"little idea at the time"; "America to Egypt"** Johannes Schmidt, "The Breeding Places of the Eel," Philosophical Transactions of the Royal Society of London, Series B 211.385 (1922), p. 181.

22 **"was considered sufficient"** Fort, *Book of Eels*, p. 95.

23 **"breeding grounds"** Schmidt, "Breeding Places of the Eel," p. 199.

23 **"known among fishes"** Johannes Schmidt, "Breeding Places and Migrations of the Eel," *Nature* 111.2776 (13 January 1923), p. 54.

26 **"somewhat humiliating"** Jacoby, "Eel Question," quoted in Schweid, *Eel*, p. 15.

CHAPTER 2. BEAVER

29 **"very gentle animal"** W. B. Clark, *A Medieval Book of Beasts: The Second-Family Bestiary: Commentary, Art, Text and Translation* (Suffolk: Boydell and Brewer, 2006), p. 130.

30 **"retire in confusion"** Anne Clark, *Beasts and Bawdy* (London: Dent, 1975), p. 92.

31 **"gentle and meek"; "great hatred"; "the most chaste"** Edward Topsell, *The History of Four Footed Beasts and Serpents and Insects* (London: DaCapo, 1967; f.p. 1658), p. 90.

32 **"dogs should chase"** Gerald of Wales, *The Itinerary of Archbishop Baldwin through Wales*, vol. 2, ed. by Sir Richard Colt Hoare (London: William Miller, 1806), p. 51.

32 **"read of the beaver"** Jean Paul Richter (ed.), *The Notebooks of Leonardo da Vinci: Compiled and Edited from the Original Manuscripts*, vol. 2 (Mineola, NY: Dover Publications, 1967), p. 1222.

32 **"bite off their Pizzles"** John Ogilby, *America: Being an Accurate Description of the New World* (London: Printed by the Author, 1671), p. 173.

32 **"vulgar errors"** Thomas Browne, *Pseudodoxia Epidemica* (London: Edward Dodd, 1646), p. iv.

33 **"Determinators of Truth"** ibid., p. 147.

33 **"clear and warrantable body"** Reid Barbour and Claire Preston (eds.), *Sir Thomas Browne: The World Proposed* (Oxford: Oxford University Press, 2008), p. 23.

33   **"repugnant to the course of nature"** Browne, quoted in *The Adventures of Thomas Browne in the Twenty-First Century*, Hugh Aldersey-Williams (London: Granta, 2015), p. 102.

34   **"beavers will tuck away"; "keeping their own treasures"** quoted in Gregory McNamee, *Aelian's on the Nature of Animals* (Dublin: Trinity University Press, 2011), p. 65.

34   **"cods"** This is Browne's term. *Pseudodoxia Epidemica*, p. 145.

34   **"eunachate"; "fruitless attempt"; "hazardous . . . if at all attempted"** Browne, ibid., p. 145.

34   **"hallucination"; "electricity"; "carnivorous"; "misconception"; "retromingent"** quoted in Hugh Aldersey-Williams, *The Adventures of Sir Thomas Browne in the Twenty-First Century*, pp. 10–12.

34   **"so named from being castrated"** Stephen A. Barney, W. J. Lewis, J. A. Beach and Oliver Berghof (eds.), *The Etymologies of Isidore of Seville* (Cambridge: Cambridge University Press, 2006), p. 21.

34   **Sanskrit word *kasturi*** Rachel Poliquin, *Beaver* (London: Reaktion, 2015), p. 58.

35   **overdose of figs** ibid., p. 57.

35   **"Testicles are defined"; "situation of these tumours"; "ground of this mistake"** Browne, *Pseudodoxia Epidemica*, p. 146.

35   **heady cocktail** John Redman Coxe, *The American Dispensatory* (Philadelphia: Carey & Lea, 1830), p. 172.

36   **"stones . . . strong and stinking savour"** Edward Topsell, *The History of Four-Footed Beasts and Serpents and Insects* (London: DaCapo, 1967; f.p. 1658), p. 38.

36   **Ancient and medieval pharmacopoeia** Poliquin, *Beaver*, p. 70.

37   **"Castoreum, Asses dung"** Topsell, *History of Four-Footed Beasts*, p. 39.

37   **popular prescription for castoreum** Poliquin, *Beaver*, p. 71.

37   **"there is rarely one"** Robert Gordon Latham (ed.), *The Works of Thomas Sydenham, MD*, vol. 2, trans. by Dr. Greenhill (London: Sydenham Society, 1848), p. 85.

37   **"delicate and irritable habits"** John Eberle, *A Treatise of the Materia Medica and Therapeutics*, quoted in Poliquin, *Beaver*, p. 53.

38   **oily brown anal secretions** G. A. Burdock, "Safety Assessment of Castoreum Extract as a Food Ingredient," *International Journal of Toxicology*, 26.1 (January–February 2007), https://www.ncbi.nlm.nih.gov/pubmed/17365147, pp. 51–5.

39   **"draw blood out of his nose"** Topsell, *History of Four-Footed Beasts*, p. 38.

39   **sequester the toxins** Poliquin, *Beaver*, p. 67.

39   Phenol from Scots pine ibid., p. 67.

40   "a few disagreeable eructations" William Alexander, *Experimental Essays on the Following Subjects: I. On the External Application of Antiseptics in Putrid Diseases. II. On the Doses and Effects of Medicines. III. On Diuretics and Sudorifics*, 2nd ed (London: Edward and Charles Dilly, 1770), p. 84.

40   "ill deserving of a place" ibid., p. 86.

41   "half-reasoning elephant" Frances Thurtle Jamieson, *Popular Voyages and Travels Throughout the Continents and Islands of Asia, Africa and America* (London: Whittaker, 1820), p. 419.

41   "the industry has been most vaunted" Nicolas Denys, *The Description and Natural History of the Coasts of North America (Acadia)*, vol. 2 (London: Champlain Society, 1908), p. 363.

42   "masons"; "carpenters"; "diggers"; "hod-carriers"; "without meddling"; "commanders"; "architect"; "chastises them" ibid., pp. 363–5.

43   "disabled tail"; "inspector of the disabled" quoted in Poliquin, *Beaver*, p. 126.

43   "build like architects" Oliver Goldsmith, *History of the Earth, and Animated Nature*, vol. 2 (1774), in *The Works of Oliver Goldsmith*, vol. 6 (London: J. Johnson, 1806), pp. 160–61.

43   "a kind of reasonable animal" Pierre François Xavier de Charlevoix, *Journal of a Voyage to North America*, quoted in Horace Tassie Martin, *Castorologia: Or, the History and Traditions of the Canadian Beaver* (London: E. Stanford, 1892), p. 167.

43   "Justice is everything among beavers" quoted in Poliquin, *Beaver*, p. 137.

43   "the lazy or idle kind"; "expell'd by the other Beavers"; "vagabonds" quoted in Gordon Sayre, "The Beaver as Native and a Colonist," *Canadian Review of Comparative Literature/Revue canadienne de littérature comparée* 22.3 4 (September and December 1995), pp. 670–71.

43   French Romantic François-René Poliquin, *Beaver*, p. 137.

45   "In proportion as man rises above a state of nature," Georges-Louis Leclerc, Comte de Buffon, *Histoire Naturelle*, vol. 6, trans. by William Smellie (London: T. Cadell, 1812), p. 128.

45   "their genius, withered by fear" ibid., p. 144.

45   "only subsisting monument" ibid., p. 130.

46   "melancholy"; "efforts feeble"; "gates of its prison" ibid., p. 134.

46   "balcony to receive the fresh air" ibid., p. 141.

46   "universal peace is maintained" ibid., p. 140.

47   intelligence gradually emerged" Poliquin, *Beaver*, p. 148.

49   **"the beaver thinks consciously"** Donald R. Griffin, *Animal Minds: Beyond Cognition to Consciousness* (Chicago: University of Chicago Press, 2001), p. 112.

50   **"quick and easy"** Frank Rosell, and Lixing Sun, "Use of Anal Gland Secretion to Distinguish the Two Beaver Species *Castor canadensis* and *C. fiber*," *Wildlife Biology* 5.2 (June 1999), http://digitalcommons.cwu.edu/biology/4/, p. 119.

CHAPTER 3. SLOTH

51   **"degraded species of sloths"** Georges-Louis Leclerc, Comte de Buffon, *Natural History, General and Particular*, vol. 9, ed. by William Wood (London: T. Cadell, 1749), p. 9.

51   **"stupidest animal"; "good to eat"** Gonzalo Fernández de Oviedo y Valdés, *The Natural History of the West Indies*, pp. 54–5.

51   **"awkward and slow"; "neither by threat"** ibid.

51   **"eight or nine minutes"; "mend their pace"** William Dampier, *Two Voyages to Campeachy*, in *A Collection of Voyages*, vol. 2 (London: James and John K. Apton, 1729), p. 61.

52   **"quadrupeds"; "on each small foot"; "neither the claws nor the feet"** Oviedo, *Natural History*, pp. 54–5.

53   **"shown in a circus"** Michael Goffart, *Function and Form in the Sloth* (Oxford: Pergamon Press, 1971), p. 75.

53   **"such an ugly animal"** Oviedo, *Natural History*, pp. 54–5.

54   **"this very deformed beast"; "Ape Bear"** Edward Topsell, *The History of Four-Footed Beasts and Serpents and Insects* (London: DaCapo, 1967; f.p. 1658), p. 15.

54   **"As Nature is lively, active, and exalted"; "One more defect"** Buffon, *Natural History*, vol. 9, p. 289.

54   **"imperfect sketches"; "wretchedness"** ibid., p. 290.

55   *Choloepus* **(meaning "crippled")** Richard Coniff, *Every Creeping Thing* (New York: Henry Holt, 1999), p. 47.

58   **"most numerically abundant"** John F. Eisenberg and Richard W. Thorington Jr., "A Preliminary Analysis of a Neotropical Mammal Fauna," *Biotropica* 5.3 (1973), pp. 150–61.

58   **"live off air"** Oviedo, *Natural History*, pp. 54–5.

59   **about 160 calories** Jonathan N. Pauli et al., "Arboreal Folivores Limit their Energetic Output, All the Way to Slothfulness," *American Naturalist* 188:2 (2016), pp. 196–204.

62 **"poor ill-formed creature"** Charles Waterton, *Wanderings in South America: The North-West of the United States, and the Antilles, in the Years 1812, 1816, 1820, and 1824* (London: B. Fellowes, 1828), p. 69.

63 **sloths in the wild** Niels C. Rattenborg, Bryson Voirin, Alexei L. Vyssotski, Roland W. Kays, Kamiel Spoelstra, Franz Kuemmeth, Wolfgang Heidrich and Martin Wikelski, "Sleeping Outside the Box: Electroencephalographic Measures of Sleep in Sloths Inhabiting a Rainforest," *Biology Letters* 4.4 (23 August 2008), pp. 402–5, http://rsbl.royalsociety publishing.org/content/4/4/402.

64 **"Sloths have no weapons"; "resembles withered herbs"** Buffon, *Natural History*, vol. 9, p. 290.

64 **980 individuals** J. K. Waage and R. C. Best (1985), "Arthropod Associates of Sloths" in G. G. Montgomery (ed.), *The Evolution and Ecology of Armadillos, Sloths and Vermilinguas,* (Washington, DC: Smithsonian Institution Press), 297–311.

65 **"A sloth in Paris"** William Beebe, "Three-Toed Sloth," *Zoologica*, 7.1 (25 March 1926), p. 13.

66 **"I have fired a gun"** ibid., p. 22.

66 **"attuned to this sound"** ibid., p. 36.

68 **American ecologists** Jonathan N. Pauli, Jorge E. Mendoza, Shawn A. Steffan, Cayelan C. Carey, Paul J. Weimar and M. Zachariah Peery, "A Syndrome of Mutualism Reinforces the Lifestyle of a Sloth," *Proceedings of the Royal Society B* 281.1778 (7 March 2014), http://dx.doi.org/10.1098/rspb.2013.3006

69 **"flying genitals"** Veronique Greenwood, "The Mystery of Sloth Poop: One More Reason to Love Science," *Time*, 22 January 2014, http://science.time.com/2014/01/22/the-mystery-of-sloth-poop-one-more -reason-to-love-science (accessed 9 July 2017).

69 **"algae gardens"** Pauli, Mendoza, Steffan, Carey, Weimar and Peery, "A Syndrome of Mutualism."

70 **speed dating** Henry Nicholls, *The Truth About Sloths*, BBC Earth website, www.bbc.co.uk/earth/story/20140916-the-truth-about-sloths.

CHAPTER 4. HYENA

71 **"hermaphroditic self-eating devourer"** Ernest Hemingway, *Green Hills of Africa* (New York: Scribner, 2015; f.p. 1935), p. 28.

72 **"rational"; "mixt natures"; "generated again"** Sir Walter Raleigh, *The Historie of the World* (London: Thomas Basset, 1687), p. 63.

73   **"vulgar notion"** John Bostock and H. T. Riley (eds.), *The Natural History of Pliny*, vol. 2 (London: George Bell, 1900), p. 296.

73   **"palpation of the scrotum"** Paul A. Racey and Jennifer D. Skinner, "Endocrine Aspects of Sexual Mimicry in Spotted Hyaenas *Crocuta crocuta*," *Journal of Zoology* 187.3 (March 1979), http://onlinelibrary.wiley.com/doi/10.1111/j.1469-7998.1979.tb03372.x/full, p. 317.

75   **recent study by Holekamp** S. M. Dloniak, J. A. French and K. E. Holekamp, "Rank-related maternal effects of androgens on behaviour in wild spotted hyaenas," *Nature* 440, 1190–1193 (27 April 2006), doi:10.1038/nature04540.

75   **"pendulous phallus"; "normal pseudo-scrotum"** Christine M. Drea et al., "Androgens and Masculinization of Genitalia in the Spotted Hyaena (*Crocuta crocuta*) 2: Effects of Prenatal Anti-Androgens," *Journal of Reproduction and Fertility* 113.1 (May 1998), p. 121.

77   **"a dirty brute"** T. H. White (ed.), *The Book of Beasts: Being a Translation from a Latin Bestiary of the Twelfth Century* (Madison, WI: Parallel Press, 2002; f.p. 1954), p. 31.

77   **"sepulchres of the dead"** ibid.

77   **"without pity for the living"** Mikita Brottman, *Hyena* (London: Reaktion, 2013) p. 40.

77   **"Place of Tombs"; "bristling mane"** Philip Henry Gosse, *The Romance of Natural History*, ed. by Loren Coleman (New York: Cosimo Classics, 2008; f.p. 1861), p. 42.

78   **"mysterious and awful" "rank and coarse"; "revolting habits"; "adapted to gorge"; "cordially detested"** Brottman, *Hyena*, p. 54.

79   **"All writers agree"** John Fortuné Nott, *Wild Animals Photographed and Described* (London: Sampson Low, Marston, Searle, & Rivington, 1886), p. 106.

79   **"heart is of large size"** Aristotle, *On the Parts of Animals*, trans. By W. Ogle (London: Kegan Paul, Trench, 1882), p. 70.

80   **"the hare, the deer, the mouse"; "manifestly timorous"** ibid., p. 71.

80   **"cowardly and will not fight"** E. P. Walker, *Mammals of the World*, quoted in Brottman, *Hyena*, p. 57.

81   **"violent fit of vomiting"** Georges-Louis Leclerc, Comte de Buffon, *Natural History* (abridged) (London: printed for C. and G. Kearsley, 1791), p. 182.

CHAPTER 5. VULTURE

85   **"attacks his enemies"** Georges-Louis Leclerc, Comte de Buffon, quoted in Stephen Jay Gould, *Leonardo's Mountain of Clams and the Diet of Worms: Essays on Natural History* (Cambridge, MA: Harvard University Press, 2011), p. 382.

85   **"cowardly, disgusting"** Buffon quoted in ibid., p.*382.

85   **"abomination"** Bible, Leviticus 11:13.

85   **"foretell the death"; "lines of battle"** T. H. White (ed.), *The Book of Beasts: Being a Translation from a Latin Bestiary of the Twelfth Century* (Madison, W I: Parallel Press, 2002; f.p. 1954), pp. 109–110.

86   **"In this bird the wit of smelling"** Robert Steele (ed.), *Mediaeval Lore from Bartholomew Anglicus* (London: Chatto and Windus, 1907), p. 132.

87   **"cruel, unclean"; "sense of smelling"; "two large apertures"** Oliver Goldsmith, *A History of the Earth, and Animated Nature,* vol. 4 (London: Wingrave and Collingwood, 1816), p. 83.

87   **"wonderfully bountiful"; "series of experiments"** John James Audubon, "An Account of the Habits of the Turkey Buzzard (*Vulturaura*) Particularly with the View of Exploding the Opinion Generally Entertained of Its Extraordinary Power of Smelling," *Edinburgh New Philosophical Journal* 2 (1826), p. 173.

89   **"voided itself freely"; "fodder and hay"** John James Audubon to John J. Jameson, ibid., p. 174.

89   **"extraordinary formation"; "wonderful works"** Charles Waterton, "Why the Sloth Is Slothful," quoted in *The World of Animals: A Treasury of Lore, Legend and Literature by Great Writers and Naturalists from the Fifth Century BC to the Present* (New York: Simon & Schuster, 1961), p. 221.

90   **"grieve from my heart";** Charles Waterton, *Essays on Natural History* (London: Frederick Warne, 1871), p. 244.

90   **"Nosarians"** Charles Waterton, "Essays on Natural History, Chiefly Ornithology," *Quarterly Review* 62 (1838), p. 85.

90   **"shattered olfactory parts"** Waterton, "Essays on Natural History," p. 244.

90   **"grammar is bad"** Charles Waterton, *Magazine of Natural History and Journal of Zoology, Botany, Mineralogy, Geology and Meteorology,* vol. 6 (London: Longman, Rees, Orme, Brown and Green, 1833), p. 215.

90   **"story of a rattlesnake"** ibid., p. 68.

90   **"Anti-Nosarians"** Waterton, "Essays on Natural History," p. 85.

90 **"skinned and cut open"** John Bachman, "Experiments Made on the Habits of the Vultures," quoted in Gene Waddell (ed.), *John Bachman: Selected Writings on Science, Race, and Religion* (Athens: University of Georgia Press, 2011), p. 76.

91 **"disappointed and surprised"; "proved very amusing"** John Bachman, "Retrospective Criticism: Remarks in Defence of [Mr Audubon] the Author of the [*Biography of the*] *Birds of America*," *Magazine of Natural History, and Journal of Zoology, Botany, Mineralogy, Geology and Meteorology*, vol. 7 (London: Longman, Rees, Orme, Brown, and Green, 1834), p. 168.

91 **"medical gentlemen"** ibid.

91 **"free from the pain"** Bachman, "Retrospective Criticism," p. 169.

91 **"offensive to the neighbours"; "sense of sight"** quoted in Waddell (ed.), *John Bachman*, p. 77.

91 **"American vulture!"** Waterton, "Essays on Natural History," p. 262.

92 **"stark, staring mad"** Ruthven Deane and William Swainson, "William Swainson to John James Audubon (A Hitherto Unpublished Letter)," *The Auk* 22.3 (July 1905), p. 251.

93 **"food-finding sense"** Herbert H. Beck, "The Occult Senses in Birds," *The Auk* 37 (1920), p. 56.

95 **"can't stop it"; "entire corpse"** quoted in David Crossland, "Police Train Vultures to Find Human Remains," *The National*, 8 January 2010, http://www.thenational.ae./news/world/europe/police-train-vultures-to-find-human-remains (accessed 12 June 2017).

96 **Miss Marple and Columbo** Michael Fröhlingsdorf, "Vulture Detective Trail Hits Headwinds," *Der Spiegel*, 28 June 2011, http://www.spiegel.de/international/germany/bird-brained-idea-vulture-detective-training-hits-headwinds-a-770994.html (accessed 9 July 2017).

97 **"They're disgusting"** quoted in Darryl Fears, "Birds of a Feather, Disgusting Together: Vultures Are Wintering Locally," *Washington Post*, 16 January 2011, https://www.washingtonpost.com/local/birds-of-a-feather-disgusting-together-vultures-are-wintering-locally/2011/01/15/AB9oNfD_story.html?utm_term=.25c80af9dd9f (accessed 12 June 2017).

97 **"smell like ammonia"; "ugly as ****"** quoted in T. Edward Nickens, "Vultures Take Over Suburbia," *Audubon*, November–December 2008, http://www.audubon.org/magazine/november-december-2008/vultures-take-over-suburbia (accessed 12 June 2017).

98 **"vomited on my son"; "ground beef"** quoted in Fears, "Birds of a Feather, Disgusting Together."

98   **"filth and the voraciousness"** Georges-Louis Leclerc, Comte de Buf-
     fon, *The Natural History of Birds* (Cambridge: Cambridge University
     Press, 2010; f.p. 1793), p. 105.

98   **"disgusting bird"** quoted in Clifford B. Frith, *Charles Darwin's Life with
     Birds: His Complete Ornithology* (Oxford: Oxford University Press, 2016), p. 44.

100 **"unprovoked rage"** Buffon, *Natural History of Birds*, p. 105.

101 **"revolting sight"** M. J. Nicoll, *Handlist of the Birds of Egypt* (Cairo: Min-
     istry of Public Works, 1919).

104 **"frogs, turtles, snakes"** quoted in Jeff Rice, "Bird Plus Plane Equals
     Snarge," *Wired*, 23 September 2005, http://archive.wired.come/science
     /discoveries/news/2005/09/68937 (accessed 12 June 2017).

104 **recent study observed a Rüppell's griffon** C. J. Kendall, M. Z. Virani,
     J. G. C. Hopcraft, K. L. Bildstein, D. I. Rubenstein (2013) "African Vul-
     tures Don't Follow Migratory Herds: Scavenger Habitat Use Is Not Me-
     diated by Prey Abundance," *PLOS ONE* 9(1): e83470, doi:10.1371/journal
     .pone.0083470.

105 **"dead camels"** quoted in Matthew Kalman, "Meet Operative PP0277:
     A Secret Agent—or Just a Vulture Hungry for Dead Camel?," *Indepen-
     dent*, 8 December 2012, http://www.independent.co.uk/news/world/
     middle-east/meet-operative-pp0277-a-secret-agent-or-just-a-vulture
     -hungry-for-dead-camel-8393578.html (accessed 12 June 2017).

CHAPTER 6. BAT

107 **"like the bat"** Georges-Louis Leclerc, Comte de Buffon, *Barr's Buffon:
     Buffon's Natural History*, vol. 6 (London: Printed for the Proprietor, 1797;
     f. p. 1749–1778), p. 239.

108 **sane British people** Charlotte-Anne Chivers, "Why Isn't Everyone
     Batty About Bats?" *Bat News*, winter edition (10) 2015.

108 **"disgusting"; "rat with leather wings"** Louis C.K., "So I Called the
     Batman . . . ," Live at the Comedy Store, 17 August 2015, https://www
     .youtube.com/watch?v=O4Eyvd T TnWY (accessed 12 June 2017).

109 **"blood-related"** Divus Basilius, quoted in Glover M. Allen, *Bats: Biol-
     ogy, Behavior, and Folklore* (Mineola, NY: Dover Publications, 2004).

110 **"imperfect quadruped"** Leclerc, *Barr's Buffon*, p. 239.

110 **"pendulous and loose"; "peculiar to man"** Georges-Louis Leclerc,
     Comte de Buffon, *A Natural History of Quadrupeds*, 3 vols, vol. 1 (Edin-
     burgh: Thomas Nelson, 1830), p. 368.

112 **"discuss the functions"** Libiao Zhang quoted in Charles Q. Choi, "Surprising Sex Behavior Found in Bats" (Live Science, 2009), http://www.livescience.com/9754-surprising-sex-behavior-bats.html (accessed 8 May 2017).

112 **"male's tongue enters"** Jayabalan Maruthupandian and Ganapathy Marimuthu, "Cunnilingus Apparently Increases Duration of Copulation in the Indian Flying Fox (Pteropus giganteus)," *PLOS ONE* 8.3 (27 March 2013), p. e59743, https://doi.org/10.1371/journal.pone.0059743.

113 *Speckmaus* Allen, *Bats*, p. 8.

114 **"great amount of blood"** Gonzalo Fernández de Oviedo y Valdés, *General and Natural History of the Indies*, quoted in Michael P. Branch (ed.), *Reading the Roots: American Nature Writing Before Walden* (Athens: University of Georgia Press, 2004; f.p. 1535), pp. 23–4.

114 **"plague of bats"** Juan Francisco Molina Solis, *Historia del Descubrimiento y Conquista del Yucatán*, vol. 3 (Merida de Yucatan: 1943), p. 38.

114 **The word *vampir*** Gary F. McCracken, "Bats and Vampires," *Bat Conservation International* 11.3 (Fall 1993), http://batcon.org/resources/media-education/bats-magazine/bat_article/603 [accessed: 12.6.2017].

115 **"vampire epidemics"** ibid.

115 **"blood from the sleeping"** Carl Linnaeus, *Systema Naturae*, tenth edition (Stockholm: Salvius, 1758), p. 31.

115 **"bloodsucker most cruel"** Johann Baptist von Spix, *Simiarum et Vespertilionum Brasiliensium Species Novae* [*New Species of Brazilian Monkeys and Bats*] (Munich: F. S. Hübschmann, 1823). BL General Reference Collection: 1899, p. 22.

115 **"long-tongued vampire"** quoted in *Blood Suckers Most Cruel*, Kevin Dodd.

115 **"hover like phantoms"** Johann Baptist von Spix, *Travels in Brazil in the Years 1817–1820*, vol. 1 (London: Longman, Hurst, Rees, Orme, Brown and Green, 1827), p. 249.

116 **"vulgar, false"** Félix de Azara, *The Natural History of the Quadrupeds of Paraguay and the River la Plata* (Edinburgh: A. & C. Black, 1838), p. xxv.

116 **"Vampyre"; "first living specimen"; "Vampyre Bat"; "his appearance"; "docile"; "appeared fond"; "devour cherries"** J. Timbs (ed.), *The Literary World: A Journal of Popular Information and Entertainment* 18 (27 July 1839), p. 274.

117 **"formidable species"; "with all the art"** Mary Trimmer, *Natural History of the Most Remarkable Quadrupeds, Birds, Fishes, Serpents, Reptiles and Insects*, vol. 1 (Chiswick: Whittingham, 1825), p. 120.

119 **"crowds of bats"** Gary McCracken, "Bats in Magic, Potions, and Medicinal Preparation," *Bat Conservation International* 10.3 (Fall 1992), http://www.batcon.org/resources/media-education/bats-magazine/bat_article/546 (accessed 8 May 2017).

119 **"wool of bat"** William Shakespeare, *Macbeth*, act 6, sc 1, l. 1560.

119 **"flying ointment"** Clive Harper, "The Witches' Flying-Ointment," *Folklore* 88.1 (1977), p. 105.

120 **"sixth sense"** quoted in Robert Galambos, "The Avoidance of Obstacles by Flying Bats: Spallanzani's Ideas (1794) and Later Theories," *Isis* 34.2 (1942), p. 138.

121 **"blind a bat"; "burning the cornea"** quoted in Donald R. Griffin, *Listening in the Dark: The Acoustic Orientation of Bats and Men* (New Haven, CT: Yale University Press, 1958), p. 59.

121 **"pair of scissors"; "thrown into the air"** quoted in Sven Dijkgraaf, "Spallanzani's Unpublished Experiments on the Sensory Basis of Object Perception in Bats," *Isis* 51.1 (1960), p. 13.

122 **"streets of a city"; "changes through their skin"** Galambos, "The Avoidance of Obstacles," p. 133.

122 **"blinded bat"; "regained its vigour"; "coat of varnish"** ibid., p. 134.

122 **"stopped up the nostrils"; "fell to the ground"; "fragments of sponge"; "flew as freely as ever"; "removal of the tongue"** quoted in Carter Beard, "Some South American Animals," *Frank Leslie's Popular Monthly* (1892), pp. 378–9.

122 **"red hot shoe nails"; "fell down"** Lazzaro Spallanzani, "Observations on the Organs of Vision in Bats," *Tillich's Philosophical Magazine* 1 (1798), p. 135.

123 **"But how, if God love me"** quoted in Griffin, *Listening in the Dark*, p. 61.

123 **"judge the distances"** quoted in Dijkgraaf, "Spallanzani's Unpublished Experiments," pp. 9–20.

123 **"organs of touch"** quoted in Griffin, *Listening in the Dark*, p. 63.

123 **"acuteness of hearing"; "see with their ears"** Galambos, "Avoidance of Obstacles," p. 137.

124 **"the *Titanic*"; "end of its tether?"; "four hours"** "A Sixth Sense for Vessels," http://chroniclingamerica.loc.gov/lccn/sn88064176/1912-09-28/ed-1/seq-10.pdf (accessed June 12, 2017).

126 **"man is not a nut"; "wild idea"** Jack Couffer, *Bat Bomb: World War II's Secret Weapon* (Austin: University of Texas Press, 1992), p. 5.

126 **"frighten"; "lowest form"** ibid.

126 **"practical"; "Japanese pest"; "used against us"** ibid., p. 6.

CHAPTER 7. FROG

129   **"singular thing"** John Bostock and H. T. Riley (eds.), *The Natural History of Pliny*, vol. 2 (London: Henry G. Bohn, 1855), pp. 462–3.

130   **"thousands of millions"** quoted in Pete Oxford and Renée Bish, "In the Land of Giant Frogs: Scientists Strive to Keep the World's Largest Aquatic Frog Off a Growing Global List of Fleeting Amphibians," 1 October 2003, https://www.nwf.org/News-and-Magazines/National-Wildlife/Animals/Archives/2003/In-the-Land-of-Giant-Frogs.aspx (accessed 20 May 2017).

134   **"old wax"; "slime of vinegar"; "damp dust"; "books"** Eugene S. McCartney, "Spontaneous Generation and Kindred Notions in Antiquity," *Transactions and Proceedings of the American Philological Association*, 51 (1920), p. 105.

134   **"fumes from the basil"; "veritable scorpions"** *Les Oeuvres de Jean-Baptiste Van Helmont*, vol. 66, trans. by Jean Le Conte (Lyon: Chez Jean Antoine Huguetan, 1670), pp. 103–9.

134   **"unclean woman"** Bondeson, *Feejee Mermaid*, p. 199.

134   **"procreated of putrefectation!"** quoted ibid., p. 200.

135   **"generated from a duck"; "no result"; "profound writer"; "too credulous"** Francesco Redi, *Experiments on the Generation of Insects* (Chicago: Open Court Publishing Company, 1909), p. 64.

135   **"raw and cooked flesh"** ibid., p. 32.

135   **"Having considered"** ibid., p. 33.

136   **"snake, some fish, some eels"** ibid.

137   **"any living body"; "offensive putrid mass"** quoted in John Waller, *Leaps in the Dark: The Making of Scientific Reputations* (Oxford: Oxford University Press, 2004), p. 42.

137   **"pair of pants"; "soft and floppy"; "adequately covered"** quoted in Mary Terrall, "Frogs on the Mantelpiece: The Practice of Observation in Daily Life," in Lorraine Daston and Elizabeth Lunbeck (eds.), *Histories of Scientific Observation* (Chicago: University of Chicago Press, 2011), p. 189.

138   **"having made the pants"** quoted in ibid.

139   **"idea of breeches"** Waller, *Leaps in the Dark*, p. 43.

140   **"godsend"** Lancelot Thomas Hogben, *Lancelot Hogben, Scientific Humanist: An Unauthorised Autobiography* (London: Merlin Press, 1998), p. 101.

143   **spectacular loss** Claude Gascon, James P. Collins, Robin D. Moore, Don R. Church, Jeanne E. McKay and Joseph R. Mendelson III (eds.),

Amphibian Conservation Action Plan (Cambridge: IUCN/SSC Amphibian Specialist Group, 2007), http://www.amphibianark.org/pdf/ACAP.pdf (accessed 12 June 2017).

147 **"plague your whole country"** Bible (English Standard Version), Exodus 8:1–4.

CHAPTER 8. STORK

149 **"diverse sorts of fowl"** Charles Morton, "An Essay into the Probable Solution of this Question: Whence Comes the Stork," quoted in Thomas Park (ed.), *The Harleian Miscellany: A Collection of Scarce, Curious, and Entertaining Pamphlets and Tracts*, vol. 5 (London: John White and John Murray, 1810), p. 506.

149 **"hands of an African"** Ragnar K. Kinzelbach, *Das Buch Vom Pfeilstorch* (Berlin: Basilisken-Presse, 2005), p. 12.

150 **"ladies glove"** quoted in Vaughan, *Wings and Rings*, p. 109.

152 **"not a fairy tale"; "If it were"** quoted in Gregory McNamee, *Aelian's on the Nature of Animals* (Dublin: Trinity University Press, 2011), p. 40.

153 **"hated"** quoted in ibid., p. 44.

153 **"Nature produces them"; "hang down by their beaks"** Gerald of Wales, *Topographia Hibernica*, quoted in Patrick Armstrong, *The English Parson-Naturalist: A Companionship Between Science and Religion* (Leominster: Gracewing Publishing, 2000), p. 31.

154 **"very naked"; "soft down"** John Gerard, *Lancashire Folk-Lore: Illustrative of the Superstitious Beliefs and Practices, Local Customs and Usages of the People of the County Palatine* (London: Frederick Warne, 1867), p. 118.

154 **"not born of flesh"; "men of religion"** Gerald of Wales, *The Historical Works of Giraldus Cambrensis* (London: Bohn, 1863), p. 36.

154 **"into hiding"; "torpor"** Aristotle, *History of Animals in Ten Books*, vol. 8, trans. by Richard Cresswell (London: George Bell, 1878), p. 213.

155 **"the sleeping one"** "Guide to North American Birds: Common Poorwill (*Phalaenoptilus nuttallii*)," National Audubon Society, http://www.audubon.org/field-guide/bird/common-poorwill (accessed 23 May 2017).

155 **"denuded of their feathers"** Aristotle, *History of Animals*, vol. 8, p. 213.

155 **"certain that swallows"** Georges Cuvier, *The Animal Kingdom*, ed. by H. M'Murtrie (New York: Carvill, 1831), p. 396.

156 "invaluable friend"; "two little prisoners"; "anxiety and convulsions"; "state of torpidity" Charles Caldwell, *Medical & Physical Memoirs: Containing, Among Other Subjects, a Particular Enquiry into the Origin and Nature of the Late Pestilential Epidemics of the United States* (Philadelphia: Thomas and William Bradford, 1801), p. 262–3.

156 "change their stations"; "northern waters,"; "assemble together" Olaus Magnus, *The History of Northern Peoples*, quoted in *Historia de Gentibus Septentrionalibus*, trans. P. Fisher and H. Higgins (London, 1998), p. 980.

157 "what truth there is" J. Hevelius, "Promiscuous Inquiries, Chiefly about Cold," *Philosophical Transactions* 1 (1665), p. 345.

158 "swallows sink themselves" ibid., p. 350.

158 "lying in clay lumps" anon. ["A Person of Learning and piety"], *An Essay Towards the Probable Solution to This Question: Whence Come the Stork, and the Turtle, and the Crane, and the Swallow When They Know and Observe the Appointed Time of Their Coming* (London: E. Symon, 1739), p. 20.

158 "when it hath bred" Charles Morton, "An Enquiry into the Physical and Literal Sense of That Scripture," in Thomas Park (ed.), *The Harleian Miscellany*, p. 506.

158 "cheerfulness" ibid., p. 506.

159 "undiscovered satellite" Cotton Mather, *The Philosophical Transactions and Collections: Abridged and Disposed Under General Heads*, vol. 5 (London: Thomas Bennet, 1721), p. 161.

159 "same line of direction" Morton, "An Enquiry," p. 510.

160 "not in Holland" Nicholaas Witsen, Emily O'Gorman and Edward Mellilo (eds.), *Beattie's Eco-Cultural Networks and the British Empire: New Views on Environmental History* (London: Bloomsbury, 2016), p. 95.

161 "always so fatigued" Daines Barrington, *Miscellanies* (London: Nichols, 1781), p. 199.

161 "highly improbable" ibid., p. 219.

161 "destitute of proof" ibid., p. 176.

162 "experimental bird" quoted in Richard Vaughan, *Wings and Rings: A History of Bird Migration Studies in Europe* (Penryn: Isabelline Books, 2009), p. 108.

162 "vain scientific humbug"; "mass murdering of storks" quoted in Raf de Bont, *Stations in the Field: A History of Place-Based Animal Research, 1870–1930* (Chicago: University of Chicago Press, 2015), p. 159.

163 "heavenly origin"; quoted in Witsen et al. (eds.), *Beattie's Eco-Cultural Networks and the British Empire*, p. 103.

163 **"our dear storks"** ibid.

165 **"sagacious birds"; "christian roof!"** quoted in Charles MacFarlane, *Constantinople in 1828: A Residence of Sixteen Months in the Turkish Capital*, vol. 1 (London: Saunders and Otley, 1829) p. 284.

166 **"vulgar errors" "a petty conceit"** Thomas Browne, quoted in Aldersey-Williams, *The Adventures of Sir Thomas Browne in the Twenty-First Century*, p. 104.

CHAPTER 9. HIPPOPOTAMUS

171 **"five cubites high"** Edward Topsell, *The History of Four-Footed Beasts and Serpents and Insects* (London: DaCapo, 1967; f.p. 1658), p. 61.

171 **"river horse"; "vomit fire"** ibid.

171 **"opening its nostrils"** quoted in David J. A. Clines, *Job 38–42: World Bible Commentary*, vol. 18B (Thomas Nelson, 2011), p. 1196.

172 **"Under the lotus"** Bible, Job 40:21.

172 **"become too bulky"** John Bostock and Henry T. Riley (eds.), *The Natural History of Pliny*, vol. 2 (London: Henry G. Bohn, 1855), p. 291.

172 **"practice of letting blood"** ibid.

175 **"so shocking"** Richard Dawkins, *The Ancestor's Tale: A Pilgrimage to the Dawn of Life* (London: Weidenfeld & Nicolson, 2010), p. 203.

178 **"animal has been celebrated"** Georges-Louis Leclerc, Comte de Buffon, *Barr's Buffon: Buffon's Natural History*, vol. 6 (London: Printed for the Proprietor, 1797; f.p. 1749–1788), p. 60.

179 **"swims well"** ibid., p. 62.

180 **"teeth are very strong"** ibid., p. 61.

180 **"powerfully armed"** ibid., p. 62.

180 **"rivers of Africa"** ibid., p. 63.

182 **"captured three"** William Kremer, "Pablo Escobar's Hippos: A Growing Problem," BBC News, 26 June 2014, http://www.bbc.co.uk/news/magazine.27905743 (accessed 28 May 2017).

184 **"highly mobile"** Chris Walzer quoted in "Moving testicles frustrate effort to calm hippos by castration," Michael Parker, *The Conversation*, 2 January 2014, https://theconversation.com/moving-testicles-frustrate-effort-to-calm-hippos-by-castration-21710.

CHAPTER 10. MOOSE

187 **"Beast Ellend"** Edward Topsell, *The History of Four-Footed Beasts and Serpent and Insects* (London: DaCapo, 1967; f.p. 1658), p. 167.

187 **"wretched case"; "joints in their legs"; "down on the ground"** ibid., p. 167.

188 **"legs without joints"** Hans-Friedrich Mueller (ed.), *Caesar: Selections from His Commentarii de Bello Gallico—Texts, Notes, Vocabulary* (Mundelein, IL: Bolchazy-Carducci, 2012), p. 242.

189 **"antelope"; "incomparable celerity"** T. H. White (ed.), *The Book of Beasts: Being a Translation from a Latin Bestiary of the Twelfth Century* (Madison, WI: Parallel Press, 2002; f.p. 1954), p. 18.

191 **"serial killer"** quoted in "Caution Warned After Alaska Moose Attacks," Associated Press, 7 May 2011, http://www.cbsnews.com/news/caution-warned-after-alaska-moose-attacks/ (accessed 24 June 2017).

191 **"drunken fashion"** Andrew Haynes, "The Animal World Has Its Junkies Too," *Pharmaceutical Journal*, 17 December 2010, http://www.pharmaceutical-journal.com/opinion/comment/the-animal-world-has-its-junkies-too/11052360.article (accessed 24 June 2017).

191 **"eating apples"** quoted in David Landes, "Swede Shocked by Backyard Elk 'Threesome,'" *The Local*, 27 October 2011, https://www.thelocal.se/20111027/36994 (accessed 24 June 2017).

192 **"extremely rare"; "several males"; "quite normal"** quoted in ibid.

192 **"big trees"** ibid.

193 **"fleshly vices"; "drunkenness and lust"; "the shrub Booze"** ibid., p. 19.

193 **"become quite tipsy"** William Drummond, *The Large Game and Natural History of South and South-East Africa* (Edinburgh: Edmonston and Douglas, 1875), p. 214.

195 **"what they were seeing"** quoted by Ronald K. Siegel in *Intoxication: The Universal Drive for Mind-Altering Substances* (Park Street Press, 1989), p. 13.

195 **"alcohol use"; "inappropriate behaviours"** Ronald K. Siegel and Mark Brodie, "Alcohol Self-Administration by Elephants," *Bulletin of the Psychonomic Society* 22.1 (July 1984), https://link.springer.com/article/10.3758/BF03333758, p. 50.

195 **"trained elephants"** Siegel, *Intoxication*, p. 120.

195 **"life-threatening clash"; "known better"** ibid., p. 122.

195 **"environmental stress"** Siegel and Brodie, "Alcohol Self-Administration by Elephants," p. 52.

195 **"favour of inebriation"; "become drunk"** Steve Morris, David Humphreys and Dan Reynolds, "Myth, Marula, and Elephant: An Assessment of Voluntary Ethanol Intoxication of the African Elephant (*Loxodonta africana*) Following Feeding on the Fruit of the Marula Tree (*Sclerocar ya birrea*)," *Physiological and Biochemical Zoology* 79.2 (March/April 2006), https://www.ncbi.nlm. nih.gov/pubmed/16555195.

196 **"drunken elephants"** quoted in Nicholas Bakalar, "Elephants Drunk in the Wild? Scientists Put the Myth to the Test," *National Geographic News*, 19 December 2005, http://news.national geographic.com/news /2005/12/1219_0 51219_drunk_elephant.html (accessed 25 June 2017).

196 **severe depression** Deer Industry Association of Australia, "Fact Sheet," https://www.deerfarming.com.au/diaa-fact-sheets (accessed 24 June 2017).

196 **"runs about, dances"** quoted in Adam Mosley, *Bearing the Heavens: Tycho Brahe and the Astronomical Community of the Late Sixteenth Century* (Cambridge: Cambridge University Press, 2007), p. 109.

197 **"In America . . . animated"; "animals are much smaller"** Georges-Louis Leclerc, Comte de Buffon, *The Natural History of Quadrupeds*, 3 vols, vol. 2 (Edinburgh: Thomas Nelson and Peter Brown, 1830), p. 31.

197 **"degenerate"** ibid., p. 51.

197 **"No American animal"** ibid., p. 31.

197 **"moral certitude"** quoted in Lee Alan Dugatkin, *Mr Jefferson and the Giant Moose: Natural History in Early America* (Chicago: University of Chicago Press, 2009), p. 35.

198 **"wallow in the mire"** Buffon, *Natural History of Quadrupeds*, p. 43.

198 **"absolutely dumb"; "less juicy"** quoted in Dugatkin, *Mr Jefferson and the Giant Moose*, p. 23.

198 **"ardour"; "small and feeble" "shrink and diminish"** Buffon, *Natural History of Quadrupeds*, p. 39. 247.

200 **"anus and the vulva"; "certainly contradicts"** James Madison to Thomas Jefferson, 19 June 1786, in *The Writings of James Madison*, ed. by Gaillard Hunt (New York: Putnam, 1900–1910), https://cdn.loc.gov/ service/mss/mjm/02/02_0677_0679.pdf (accessed 24 June 2017).

200 **"When Mr Jefferson"** quoted in Paul Ford (ed.), *The Works of Thomas Jefferson; Correspondence and Papers, 1816–1826*, vol. 7 (New York: Cosimo Books, 2009), p. 393.

200 **"absolutely unacquainted"; "reindeer could walk under"** quoted in ibid., p. 393.

200 **"horns one foot long"** quoted in ibid., p. 393.

200 **"rattling sound"** quoted in Dugatkin, *Mr Jefferson and the Giant Moose*, p. 107.

200 **"seven to ten feet tall"; "extraordinary size"** quoted in ibid., p. 91.

200 **"The readiness with which you undertook"** Thomas Jefferson to John Sullivan, 7 January 1786, Founders Archive, https://founders.archives.gov/documents/Jefferson/01-09-02-0145 (accessed 24 June 2017).

201 **"bones of the head"; "by sewing"** ibid.

201 **"state of putrefaction"; "horns of this moose"** John Sullivan to Jefferson, 16 April 1787, Founders Archive, https://founders.archives.gov/documents/Jefferson/01-11-02-0285 (accessed 24 June 2017).

201 **"remarkably small"; "five or six times"** Thomas Jefferson to Georges-Louis Leclerc, Comte de Buffon, 1 October 1787, American History, http://www.let.rug.nl/usa/presidents/thomas-jefferson/letters-of-thomas-jefferson/jef163.php (accessed 24 June 2017).

201 **"set these things right"** quoted in Ford (ed.), *Works of Thomas Jefferson*, p. 394.

CHAPTER 11. PANDA

203 **"bad at sex"** "Pandanomics," *The Economist*, 18 January 2014, http://www.economist.com/news/united-states/21594315-costly-bumbling-washington-has-perfect-mascot-pandanomics (accessed 11 May 2017).

203 **"not a strong species"** Chris Packham, "Let Pandas Die," *Radio Times*, 22 November 2009, http://www.radiotimes.com/news/2009-09-22/chris-packham-let-pandas-die (accessed 7 July 2017).

203 **two pandas**: Henry Nicholls, "The Truth About Giant Pandas," BBC website, www.bbc.co.uk/earth/story/20150310-the-truth-about-giant-pandas.

204 **"It is unbelievable"** quoted in Richard Conniff, *The Species Seekers: Heroes, Fools, and the Mad Pursuit of Life on Earth* (New York: W. W. Norton, 2010), p. 317.

204 **"barks like a dog"** quoted in ibid., p. 307.

204 **"excellent black and white bear"; "novelty to science"** quoted in Henry Nicholls, *Way of the Panda: The Curious History of China's Political Animal* (London: Profile Books, 2011), p. 9.

204 **"not look very fierce"; "full of leaves"** quoted in Conniff, *Species Seekers*, p. 315.

205 **"most closely related"** George Schaller, *The Last Panda* (Chicago: University of Chicago Press, 1994), p. 266.

205 **"panda is a panda"; "hope there is a yeti"** ibid., p. 262.

206 **"misshapen lump"** quoted in Gregory McNamee, *Aelian's on the Nature of Animals* (Dublin: Trinity University Press, 2011), p. 26.

206 **"licks it into the form"** ibid., p. 59.

206 **"dehydrate and atrophy"; "makes her fart"; "eats a bunch of ants"; "enjoys a fine defecation"** ibid., p. 60.

208 **Boaty McBoatface** Hannah Ellis-Petersen, "Boaty McBoatface Wins Poll to Name Polar Research Vessel," *Guardian*, 17 April 2016, https://www.theguardian.com/environment/2016/apr/17/boaty-mcboatface-wins-poll-to-name-polar-research-vessel (accessed 8 July 2017).

209 **"Bronx officials"** Ramona Morris and Desmond Morris, *Men and Pandas* (London: Hutchinson and Co., 1966), p. 92.

209 **"raised her tail"; "intense embarrassment"; "somewhat warped"; "complete strangers"** Oliver Graham-Jones, *Zoo Doctor* (Fontana Books, 1973), p. 140.

209 **"somewhat warped"; "complete strangers"** ibid., p. 140.

210 **"Chi-Chi's long isolation"** ibid., p. 141.

211 **George Schaller was the first** George B. Schaller, Hu Jinchu, Pan Wenshi and Zhu Jing, *The Giant Pandas of Wolong* (Chicago: University of Chicago Press, 1985).

212 **"spermatozoa"** Susie Ellis, Anju Zhang, Hemin Zhang, Jinguo Zhang, Zhihe Zhang, Mabel Lam, Mark Edwards, JoGayle Howard, Donald Janssen, Eric Miller and David Wildt, "Biomedical Survey of Captive Giant Pandas: A Catalyst for Conservation Partnerships in China," in Donald Lindburg and Karen Baragona (eds.), *Giant Pandas: Biology and Conservation* (Berkeley: University of California Press, 2004), p. 258, http://www.jstor.org/stable/10.1525/j.ctt1ppskn.

212 **"squat," "legcock"; "handstand"** Angela M. White, Ronald R. Swaisgood, Hemin Zhang, "The Highs and Lows of Chemical Communication in Giant Pandas (*Ailuropoda melanoleuca*): Effect of Scent Deposition Height on Signal Discrimination," *Behavioural Ecology Sociobiology* 51.6 (May 2002), pp. 519–29, https://link.springer.com/article/10.1007/s00265-002-0473-3 (accessed 22 June 2017).

214 **"George's girlfriend"** Henry Nicholls, *Lonesome George: The Life and Loves of a Conservation Icon* (New York: Palgrave, 2007), p. 30.

216 **"taking off the pants"** quoted in Lijia Zhang, "Edinburgh Zoo's Pandas Are a Big Cuddly Waste of Money," *Guardian*, 7 December 2011, https://www.theguardian.com/commentisfree/2011/dec/07/edinburgh-zoo-pandas-big-waste-money (accessed 11 May 2017).

218 **$3.5 billion** Kathleen C. Buckingham, Jonathan Neil, William David and Paul R. Jepson, "Diplomats and Refugees: Panda Diplomacy, Soft 'Cuddly' Power, and the New Trajectory in Panda Conservation," *Environmental Practice* 15.3 (2013), pp. 262–70, https://www.researchgate.net/publication/255981642.

218 **"to seal the deal"; "soft power influence"** quoted in Melissa Hogenboom, "China's New Phase of Panda Diplomacy," BBC News, 25 September 2013, http://www.bbc.co.uk/news/science-environment-24161385 (accessed 22 June 2017).

219 **"Panda-monium"** Brynn Holland, "Panda Diplomacy: The World's Cutest Ambassadors," History Channel, 16 March 2017, www.history.com/news/panda-diplomacy-the-worlds-cutest-ambassadors.

219 **"Chinese slang"** quoted in Christopher Klein, "When "Panda-Monium" Swept America," History Channel, 9 January 2014, http://www.history.com/news/when-panda-monium-swept-america (accessed 22 June 2017).

219 **"giraffe-driven craze"** Eric Ringmar, "Audience for a Giraffe: European Exceptionalism and the Quest for the Exotic," *Journal of World History* 17.4 (December 2006), http://www.jstor.org/stable/20079397, p. 385.

219 **"exclusive natural existence"** Falk Hartig, "Panda Diplomacy: The Cutest Part of China's Public Diplomacy," *Hague Journal of Diplomacy* 8.1 (2013), https://eprints.qut.edu.au/59568.

220 **"mistake of stroking them"** When Pandas Attack! (blog), https://whenpandasattack.wordpress.com (accessed 11 May 2017).

221 **study of the top bite-forces** Christiansen and Stephen Wroe (2007), "Bite force and evolutionary adaptations to feeding ecology in carnivores," *Ecology* 88, pp. 347–58, 10.1890/0012-9658(2007)88[347:BFAEAT]2.0.CO;2.

221 **"they were cute"** ibid.

CHAPTER 12. PENGUIN

223 **"a penguin"** Apsley Cherry-Garrard, *The Worst Journey in the World: Antarctic 1910–1913*, vol. 2 (New York: George H. Doran, 1922), p. 560.

225 **"fowl which could not fly"; "bigness of geese"** *Sir Francis Drake's Famous Voyage Round the World* (1577), quoted in Tui de Roy, Mark Jones and Julie Cornthwaite, *Penguins: The Ultimate Guide* (Princeton, NJ: Princeton University Press, 2014), p. 151.

226 **"You take a kettle"** Errol Fuller, *The Great Auk: The Extinction of the Original Penguin* (Piermont, NH: Bunker Hill, 2003), p. 34.

227 **"arse-feet"** quoted in Oliver Goldsmith, *A History of the Earth, and Animated Nature*, vol. 4 (Philadelphia: T. T. Ash, 1824), p. 83.

227 **"eccentric to a degree"** Edward A. Wilson, *Report on the Mammals and Birds, National Antarctic Expedition 1901–1904*, vol. 2 (London: Aves, 1907), p. 11.

227 **"have in the emperor penguin"** ibid., p. 38.

228 **"*Archaeopteryx*"; "find real teeth"** Edward A. Wilson and T. G. Taylor, *With Scott: The Silver Lining* (New York: Dodd, Mead and Company, 1916), p. 244.

230 **"point of suffering"** Cherry-Garrard, *Worst Journey*, p. 237.

230 **"tremendous row"** ibid., p. 268.

230 **"boiling oil"; "stifle his groans"** ibid., p. 273.

230 **"mistrusted that stove"** ibid., p. 274.

230 **"having a fit"; "little tiny strips"** ibid., p. 276.

230 **"real death"** ibid., p. 281.

231 **"lives had been taken away"** ibid., p. 284.

231 **"Sacred eggs"; "Who are you?"** ibid., p. 299.

231 **"Cape Crozier embryos"** Sara Wheeler, *Cherry: A Life of Apsley Cherry-Garrard* (London: Vintage, 2007), p. 186.

232 **"penguin embryology"** C. W. Parsons, "Penguin Embryos: British Antarctic Terra Nova Expedition 1910—Natural History Reports," *Zoology* 4.7 (1934), p. 253.

232 **"hard life"** Cherry-Garrard, *Worst Journey*, vol. 1, p. 269.

232 **"like children"; "own importance"** ibid., p. 50.

232 **"look of a child"** William Clayton, "An Account of Falkland Islands," *Philosophical Transactions of the Royal Society of London* 66 (1 January 1776), p. 103.

232 **"like little Children"** John Narborough, Abel Tasman, John Wood and Friderich Martens, *An Account of Several Late Voyages and Discoveries to the South and North* (Cambridge: Cambridge University Press, 2014; f.p. 1711), p. 59.

232 **"ridiculous air of gravity"** "The Zoological Gardens Regents Park," *The Times*, 18 April 1865, p. 10.

233 **"incredible true story"** Luc Jacquet and Bonne Pioche (dirs), *March of the Penguins* (National Geographic Films, 2005).

233 **"passionately affirms"; "Some of the circumstances"** quoted in Jonathan Miller, "March of the Conservatives: Penguin Film as Political

Fodder," *New York Times*, 13 September 2005, http://www. nytimes
.com/2005/09/13/science/march-of-the-conservatives-penguin-film
-as-political-fodder.html (accessed 26 June 2017).

234 **"sexually motivated"** Bruce Bagemihl, *Biological Exuberance: Animal
Homosexuality and Natural Diversity* (New York: St Martin's Press, 1999),
p. 115.

235 **"rocked the gay scene"** Andrew Sullivan quoted in Miller, "New Love
Breaks Up Six-Year Relationship at Zoo," *New York Times*, 24 September
2005.

238 **"gangs of hooligan cocks"; "beyond their control"** Douglas G. D.
Russell, William J. L. Sladen and David G. Ainley, "Dr George Murray
Levick (1876–1956): Unpublished Notes on the Sexual Habits of the Adé-
lie Penguin," *Polar Record* 48.4 (October 2012), https://doi.org/10.1017/
S0032247412000216, p. 388.

238 **"constant acts of depravity"** ibid., p. 392.

238 **"sexually misused" "very eyes of its parent"; "suffer indignity"** ibid.

239 **"have this cut out"** ibid., p. 388.

239 **"astonishing acts of depravity"** ibid., p. 389.

239 **"engaged in sodomy"; "no crime too low"** ibid.

240 **"feral pigeon"; "dead house martin"; "much smaller bird"** username
Zheljko, "Avian Necrophilia" discussion board, *Birdforum*, 6 May 2014
18:43, http://www.birdforum.net/showthread.php?t=282175 (accessed
on 23 May 2017).

240 **"presenting to a male"; "was duly squashed, its turn"** username
Farnboro John, "Avian Necrophilia" discussion board, *Birdforum*, 6 May
2014 17:20, http://www.birdforum.net/showthread.php?t=282175 (accessed on 23 May 2017).

241 **"Grim stuff"** username Capercaillie71, "Avian Necrophilia" discussion
board, *Birdforum*, 6 May 2014 21:34, http://www.birdforum.net/show
thread.php?t=282175 (accessed on 23 May 2017).

241 **"irresistible" "self-adhesive white O's"** ibid., p. 390.

241 **"males to copulate"** ibid., p. 389.

CHAPTER 13. CHIMPANZEE

243 **"A brute"** Georges-Louis Leclerc, Comte de Buffon, *History of Quadrupeds*, vol. 3 (Edinburgh: Thomas Nelson, 1830), p. 248.

245 **"constitution of the ape is hot" "habits of beasts"** Hildegard of

Bingen, quoted in H. W. Janson, *Apes and Ape Lore in the Middle Ages and the Renaissance* (London: Warburg Institute, 1952), p. 77.

246 **"two kinds of Monsters"; "pongos"; "engecos"** Andrew Battel, *Purchas, His Pilgrimage,* quoted in Robert Yerkes and Ada Yerkes, *The Great Apes: A Study of Authropoid Life* (New Haven, CT: Yale University Press, 1929), pp. 42–3.

246 **"they do not very well love"; "pernicious sort of brutes"** Willem Bosman, *A New and Accurate Description of the Coast of Guinea* (London: Alfred Jones, 1705), p. 254.

246 **"hideous countenance"; "own excrement"; "illicit sexual intercourse"** Jonathan Marks, *What It Means to Be 98% Chimpanzee: Apes, People, and Their Genes* (Berkeley: University of California Press, 2002), p. 19.

247 **"apt to think"** Edward Tyson, quoted in John M. Batcherlder, "Letters to the Editor: Dr. Edward Tyson and the Doctrine of Descent," *Science* 11.270 (1888), pp. 169–70.

247 **"cave-dwelling man"** quoted by Marks, *What It Means to Be 98% Chimpanzee,* p. 21.

248 **"in the human species"** Georges-Louis Leclerc, Comte de Buffon, *Barr's Buffon: Buffon's Natural History,* vol. 9 (London: Symonds, 1797), p. 157.

248 **"fleshy posteriors"** ibid., p.175.

248 **"power of thought"** ibid., p. 138.

248 **"superior principle"** ibid., p. 167.

248 **"supreme master"** Richard Owen, "On the Characters, Principles of Division, and Primary Groups of the Class Mammalia," *Journal of the Proceedings of the Linnean Society I: Zoology* (London: Longman, 1857), p 34.

249 **"ruling brain"** Richard Owen, quoted in Carl Zimmer, "Searching for Your Inner Chimp," *Natural History,* Dec. 2002–Jan. 2003.

249 **"what a Chimpanzee [would] say"** Charles Darwin to J. D. Hooker, 5 July 1857, Darwin Correspondence Project, http://www.darwinproject .ac.uk/DCP-LETT-2117 (accessed 5 May 2017).

249 **"not ashamed"** J. R. Lucas, "Wilberforce and Huxley: A Legendary Encounter," *Historical Journal* 22.2 (1979), pp. 313–30.

249 **"mendacious humbug"** Thomas Henry Huxley, quoted in Stephen Jay Gould, *Leonardo's Mountain of Clams and the Diet of Worms* (Cambridge, MA: Harvard University Press, 2011), p. 129.

249 **"Corinthian portico"** Thomas Henry Huxley to Joseph Dalton Hooker, 5 September 1858, in G. W. Beccaloni (ed.), Wallace Letters Online, http://www.nhm.ac.uk/research-curation/scientific-resources/

collections / library-collections / wallace-letters-online / 3758 / 3670 / T / details.html (accessed 25 June 2017).

250 **"interesting evidence"; "decisive blow"** Kirill Rossiianov, "Beyond Species: Ll'ya Ivanov and His Experiments on Cross-Breeding Humans with Anthropoid Apes," *Science in Context* 15.2 (2002), p. 279.

250 **"possible and desirable"** ibid.

251 **"rejuvenation therapy"** Serge Voronoff, *The Conquest of Life* (New York: Brentano, 1928), p. 130.

251 **"monkey gland"** ibid.

251 **"aphrodisiac"** ibid., p. 150.

252 **"not completely fresh"** Rossiianov, "Beyond Species," p. 289.

252 **"bolt from the blue"** ibid., p. 289.

255 **"I am a psychotherapist"** Maurice K. Temerlin, *Lucy: Growing up Human—a Chimpanzee Daughter in a Psychotherapist's Family* (Palo Alto, CA: Science & Behavior Books, 1975), p. 1.

255 **"symbolic equivalent"** ibid., p. 8.

255 **"Jewish momma's boy"** ibid., p. 130.

255 **"fix her a cocktail or two"** ibid., p. 49.

256 **"redefine tool"** Louis Leakey, quoted in David Quammen, "Fifty Years at Gombe," *National Geographic*, October 2010, http://ngm.national geographic.com/print/2010/10/jane-goodall/quammen-text (accessed 27 May 2017).

256 **"see what would happen"** Temerlin, *Lucy: Growing Up Human*, p. 109.

258 **"Lucy was into everything"** ibid., p. 19.

263 **"building shrines"; "sacred tree"** Simon Barnes, "Is This Proof Chimps Believe in God?," *Daily Mail*, 4 March 2006, http://www.dailymail .co.uk/sciencetech/article-3475816/Is-proof-chimps-believe-God -Scientists-baffled-footage-primates-throwing-rocks-building-shrines -sacred-tree-no-reason.html (accessed 27 May 2017).

264 **"awe and wonder"** Jane Goodall, "Waterfall Displays," Vimeo, 3 January 2011, https://vimeo.com/18404370 (accessed 27 June 2017).

265 **"They are not men"** Edward Topsell, *The History of Four-Footed Beasts and Serpents and Insects*, vol. 1 (London: DaCapo, 1967; f.p. 1658), p. 3.

Bibliography

INTRODUCTION

Aldersey-Williams, Hugh, *The Adventures of Sir Thomas Browne in the Twenty-First Century* (London: Granta, 2015)

Clark, Anne, *Beasts and Bawdy* (London: Dent, 1975)

Curley, Michael J. (trans.), *Physiologus: A Medieval Book of Natural Lore* (Chicago: University of Chicago Press, 1979)

Raven, Charles E., *English Naturalists from Neckam to Ray: A Study of the Making of the Modern World* (Cambridge: Cambridge University Press, 2010)

White, T. H., *The Book of Beasts: Being a Translation from a Latin Bestiary of the Twelfth Century* (Madison, WI: Parallel Press, 2002; f.p. 1954)

CHAPTER 1. EEL

Amilhat, Elsa, Kim Aarestrup, Elisabeth Faliex, Gaël Simon, Håkan Westerberg and David Righton, "First Evidence of European Eels Exiting the Mediterranean Sea During Their Spawning Migration," *Nature Scientific Reports* 6.21817 (24 February 2016), https://www.nature.com/articles/srep21817

Aristotle, "Historia Animalium," *The Works of Aristotle*, vol. 4, trans. by D'Arcy Wentworth Thompson (Oxford: Clarendon Press, 1910)

Cairncross, David, *The Origin of the Silver Eel: With Remarks on Bait and Fly Fishing* (London: G. Shield, 1862)

Fort, Tom, *The Book of Eels* (London: HarperCollins, 2002)

Goode, G. Brown, "The Eel Question," *Transactions of the American Fisheries Society*, vol. 10 (New York: Johnson Reprint Corp., 1881), pp. 81–124

Grassi, G. B., "The Reproduction and Metamorphosis of the Common Eel (Anguilla vulgaris)," *Reproduction and Metamorphosis of Fish* (1896), p. 371

Jacoby, Leopold, "The Eel Question," in US Commission of Fish and Fisheries, Report of the Commissioner for 1879 (Washington: US Government Printing Office, 1882), http://penbay.org/cof/COF_1879_IV.pdf

Magnus, Albert, *On Animals: A Medieval Summa Zoologica*, vol. 2, trans. by
    Kenneth F. Kitchell Jr. and Irven Michael Resnick (Baltimore: John Hop-
    kins University Press, 1999)
Marsh, M. C., "Eels and the Eel Questions," *Popular Science Monthly* 61.25
    (September 1902), pp. 426–33
Poulsen, Bo, *Global Marine Science and Carlsberg: The Golden Connections of
    Johannes Schmidt (1877–1933)* (Leiden: Brill, 2016)
Prosek, James, *Eels: An Exploration, from New Zealand to the Sargasso, of the
    World's Most Amazing and Mysterious Fish* (London: HarperCollins, 2010)
Righton, David, Kim Aarestrup, Don Jellyman, Phillipe Sébert, Guido van
    den Thillart and Katsumi Tsukamato, "The Anguilla spp. Migration
    Problem: 40 Million Years of Evolution and Two Millennia of Specula-
    tion," *Journal of Fish Biology* 81.2 (July 2012), pp. 365–86, https://www
    .ncbi.nlm.nih.gov/pubmed/22803715
Schmidt, Johannes, "Breeding Places and Migrations of the Eel," *Nature*
    111.2776 (13 January 1923), pp. 51–4
Schmidt, Johannes, "The Breeding Places of the Eel," *Philosophical Trans-
    actions of the Royal Society of London*, Series B 211.385 (1922), pp. 179–208
Schweid, Richard, *Consider the Eel: A Natural and Gastronomic History* (Chapel
    Hill: University of North Carolina Press, 2002)
Schweid, Richard, *Eel* (London: Reaktion, 2009)
Schweid, Richard, "Slippery Business: Scientists Race to Understand the Re-
    productive Biology of Freshwater Eels," *Natural History* 118.9 (November
    2009), pp. 28–33, http://www.naturalhistorymag.com/features/291856/
    slippery-business
Walton, Izaak, and Charles Cotton, *The Complete Angler: Or the Contemplative
    Man's Recreation*, ed. by John Major (London: D. Bogue, 1844)

CHAPTER 2. BEAVER

Browne, Thomas, *Pseudodoxia Epidemica* (London: Edward Dodd, 1646)
Buffon, Georges-Louis Leclerc, Comte de, *History of Quadrupeds*, vol. 6.,
    trans. by William Smellie (London: T. Cadell, 1812)
Campbell-Palmer, Róisín, Derek Gow and Robert Needham, *The Eurasian
    Beaver* (Exeter: Pelagic Publishing, 2015)
Clark, W. B., *A Medieval Book of Beasts: The Second-Family Bestiary: Commen-
    tary, Art, Text and Translation* (Suffolk: Boydell and Brewer, 2006)

Dolin, Eric Jay, *Fur, Fortune, and Empire: The Epic History of the Fur Trade in America* (New York: W. W. Norton, 2011)

Gerald of Wales, *The Itinerary of Archbishop Baldwin Through Wales*, vol. 2, ed. by Sir Richard Colt Hoare (London: William Miller, 1806)

Gould, James L., and Carol Grant Gould, *Animal Architects: Building and the Evolution of Intelligence* (New York: Basic Books, 2012)

Gould, Stephen Jay, *The Mismeasure of Man* (New York: W. W. Norton, 1996)

Griffin, Donald R., *Animal Minds: Beyond Cognition to Consciousness* (Chicago: University of Chicago Press, 2001)

McNamee, Gregory, *Aelian's on the Nature of Animals* (Dublin: Trinity University Press, 2011)

Martin, Horace Tassie, *Castorologia: Or, the History and Traditions of the Canadian Beaver* (London: E. Stanford, 1892)

Mortimer, C., "The Anatomy of a Female Beaver, and an Account of Castor Found in Her," *Philosophical Transactions* 38 (1733), pp. 172–83, http://rstl.royalsocietypublishing.org/content/38/427 435/172

Müller-Schwarze, Dietland, *The Beaver: Its Life and Impact*, 2nd ed. (Ithaca, NY: Cornell University Press, 2011)

Müller-Schwarze, Dietland and Lixing Sun, *The Beaver: History of a Wetlands Engineer* (Ithaca, NY: Cornell University Press, 2003)

Nolet, Bart A., and Frank Rosell, "Comeback of the Beaver *Castor fiber*: An Overview of Old and New Conservation Problems," *Biological Conservation* 83.2 (1998), pp. 165–73, http://hdl.handle.net/20.500.11755/6cc63738 -2516-44f4-b31a-f4d686b4e249

Platt, Carolyn V., *Creatures of Change: An Album of Ohio Animals* (Kent, OH: Kent State University Press, 1998)

Poliquin, Rachel, *Beaver* (London: Reaktion, 2015)

Sax, Boria, *The Mythical Zoo: An Encyclopedia of Animals in World Myth, Legend, and Literature* (Santa Barbara, CA : ABC-Clio, 2001)

Sayre, Gordon, "The Beaver as Native and a Colonist," *Canadian Review of Comparative Literature/Revue canadienne de littérature comparée* 22.3–4 (September and December 1995), pp. 659–82

Simon, Matt, "Fantastically Wrong: Why People Used to Think Beavers Bit Off Their Own Testicles," wired.com, 2014

Tasca, Cecilia, Mariangela Rapetti, Mauro Giovanni Carta and Bianca Fadda, "Women and Hysteria in the History of Mental Health," *Clinical Practice and Epidemiology in Mental Health* 8 (October 2012), pp. 110–19, https://www.ncbi.nlm.nih.gov/pmc/articles/PMC3480686

Wilsson, Lars, *Observations and Experiments on the Ethology of the European Beaver (Castor Fiber L.): A Study in the Development of Phylogenetically Adapted Behaviour in a Highly Specialized Mammal* (Uppsala: Almqvist & Wiksell, 1971)

CHAPTER 3. SLOTH

Beebe, William, "Three-Toed Sloth," *Zoologica*, 7.1 (25 March 1926)

Buffon, Georges-Louis Leclerc, Comte de, *Natural History, General and Particular*, vol. 9, ed. by William Wood (London: T. Cadell, 1749)

Choi, Charles Q., "Freak of Nature: Sloth Has Rib-Cage Bones in Its Neck," *LiveScience*, 21 October 2010, https://www.livescience.com/10178-freak -nature-sloth-rib-cage-bones-neck.html

Cliffe, Rebecca N., Judy A. Avey-Arroyo, Francisco J. Arroyo, Mark D. Holton and Rory P. Wilson, "Mitigating the Squash Effect: Sloths Breathe Easily Upside Down," *Biology Letters* 10.4 (April 2014), http://rsbl.royalsocietypublishing.org/content/10/4/20140172

Cliffe, Rebecca N., Ryan J. Haupt, Judy A. Avey-Arroyo and Rory P. Wilson, "Sloths Like It Hot: Ambient Temperature Modulates Food Intake in Brown-Throated Sloth (*Bradypus variegatus*)," *PeerJ* 3 (2 April 2015), p. e875, https://www.ncbi.nlm.nih.gov/pubmed/25861559

Conniff, Richard, *Every Creeping Thing: True Tales of Faintly Repulsive Wildlife* (New York: Henry Holt, 1999)

Eisenberg, John F., and Richard W. Thorington Jr., "A Preliminary Analysis of a Neotropical Mammal Fauna," *Biotropica* 5.3 (1973), pp. 150–61

Goffart, Michael, *Function and Form in the Sloth* (Oxford: Pergamon Press, 1971)

Gould, Carol Grant, *The Remarkable Life of William Beebe: Naturalist and Explorer* (Washington, DC: Island Press: 2004)

Gould, Stephen Jay, *Leonardo's Mountain of Clams and the Diet of Worms* (Cambridge, MA: Belknap Press of Harvard University Press, 2011)

Horne, Genevieve, "Sloth Fur Has a Symbiotic Relationship with Green Algae," *Biomed Central* blog, 14 April 2010, https://blogs.biomedcentral .com/on-biology/2010/04/14/sloth-fur-has-symbiotic-relationship -with-green-algae (accessed 28 May 2017)

Montgomery, G. Gene, and M. E. Sunquist, "Habitat Selection and Use by Two-Toed and Three-Toed Sloths," in *The Ecology of Arboreal Folivores* (Washington, DC: Smithsonian Institute, 1978), pp. 329–59

Oviedo y Valdés, Gonzalo Fernández de, *The Natural History of the West Indies*, ed. by Sterling A. Stoudemire (Chapel Hill: University of North Carolina Press, 1959), pp. 54–5

Pauli, Jonathan N., Jorge E. Mendoza, Shawn A. Steffan, Cayelan C. Carey, Paul J. Weimar and M. Zachariah Peery, "A Syndrome of Mutualism Reinforces the Lifestyle of a Sloth," *Proceedings of the Royal Society B* 281.1778 (7 March 2014), http://dx.doi.org/10.1098/rspb.2013.3006

Rattenborg, Niels C., Bryson Voirin, Alexei L. Vyssotski, Roland W. Kays, Kamiel Spoelstra, Franz Kuemmeth, Wolfgang Heidrich and Martin Wikelski, "Sleeping Outside the Box: Electroencephalographic Measures of Sleep in Sloths Inhabiting a Rainforest," *Biology Letters* 4.4 (23 August 2008), pp. 402–5, http://rsbl.royalsocietypublishing.org/content/4/4/402

Voirin, Bryson, Roland Kays, Martin Wikelski and Margaret Lowman, "Why Do Sloths Poop on the Ground?," in Margaret Lowman, T. Levy and Soubadra Ganesh (eds.), *Treetops at Risk* (New York: Springer, 2013), pp. 195–9

CHAPTER 4. HYENA

Aristotle, *On the Parts of Animals*, trans. by W. Ogle (London: Kegan Paul, Trench, 1882)

Baynes-Rock, Markus, *Among the Bone Eaters: Encounters with Hyenas in Harar* (State College: Pennsylvania State University Press, 2015)

Benson-Amram, Sarah, and Kay E. Holekamp, "Innovative Problem Solving by Wild Spotted Hyenas," *Proceedings of the Royal Society B* 279.1744 (October 2012), pp. 4087–95, https://www.ncbi.nlm.nih.gov/pmc/articles/PMC3427591

Benson-Amram, Sarah, Virginia K. Heinen, Sean L. Dryer and Kay E. Holekamp, "Numerical Assessment and Individual Call Discrimination by Wild Spotted Hyaenas, *Crocuta crocuta*," *Animal Behaviour* 82.4 (October 2011), pp. 743–52, https://doi.org/10.1016/j.anbehav.2011.07.004

Brottman, Mikita, *Hyena* (London: Reaktion, 2013)

Coscia, Laurence G. Frank, Paul Licht and Stephen E. Glickman, "Androgens and Masculinization of Genitalia in the Spotted Hyaena (*Crocuta crocuta*) 2: Effects of Prenatal Anti-Androgens," *Journal of Reproduction and Fertility* 113.1 (May 1998), pp. 117–27, https://www.ncbi.nlm.nih.gov/pubmed/9713384

Cunha, Gerald R., Yuzhuo Wang, Ned J. Place, Wenhui Liu, Larry Baskin and Stephen E. Glickman, "Urogenital System of the Spotted Hyena

(*Crocuta crocuta Erxleben*): A Functional Histological Study," *Journal of Morphology* 256.2 (May 2003), pp. 205–18, http://onlinelibrary.wiley.com/doi/10.1002/jmor.10085/full

Drea, Christine M., and Allisa N. Carter, "Cooperative Problem Solving in a Social Carnivore," *Animal Behaviour* 78.4 (October 2009), pp. 967–77, http://dx.doi.org/10.1016/j.anbehav.2009.06.030

Frank, Laurence G., "Evolution of Genital Masculinization: Why do Female Hyaenas Have Such a Large 'Penis'?," *Trends in Ecology & Evolution* 12.2 (February 1997), pp. 58–62, https://www.ncbi.nlm.nih.gov/pubmed/21237973

Frank, Laurence G., Stephen E. Glickman and Irene Powch, "Sexual Dimorphism in the Spotted Hyaena (*Crocuta crocuta*)," *Journal of Zoology* 221.2 (1990), pp. 308–13, http://onlinelibrary.wiley.com/doi/10.1111/j.1469-7998.1990.tb04001.x/full

Glickman, Stephen E., "The Spotted Hyena from Aristotle to *The Lion King*: Reputation Is Everything," *Social Research* 62.3 (Fall 1995), pp. 501–37

Glickman, Stephen E., Gerald R. Cunha, Christine M. Drea, Al J. Conley and Ned J. Place, "Mammalian Sexual Differentiation: Lessons from the Spotted Hyena," *Trends in Endocrinology & Metabolism* 17.9 (November 2006), pp. 349–56, https://www.ncbi.nlm.nih.gov/pubmed/17010637

Gould, Stephen Jay, *Hen's Teeth and Horse's Toes: Further Reflections in Natural History* (New York: W. W. Norton, 1984)

Holekamp, Kay E., Sharleem Sakai and Barbara Lundrigan, "Social Intelligence in the Spotted Hyena (*Crocuta crocuta*)," *Philosophical Transactions of the Royal Society of London B* 362.1480 (29 April 2007), pp. 523–38, https://www.ncbi.nlm.nih.gov/pmc/articles/PMC2346515

Hyaena Specialist Group, www.hyaenaspecialistgroup.org

Kemper, Steve, "Who's Laughing Now?," *Smithsonian Magazine*, May 2008.

Kruuk, Hans, *The Spotted Hyena: A Study of Predation and Social Behaviour* (Chicago: University of Chicago Press, 1972)

Nicholls, Henry, "The Truth About Spotted Hyenas," BBC Earth, 28 October 2014, http://www.bbc.co.uk/earth/story/20141028-the-truth-about-spotted-hyenas

Racey, Paul A., and Jennifer D. Skinner, "Endocrine Aspects of Sexual Mimicry in Spotted Hyaenas *Crocuta crocuta*," *Journal of Zoology* 187.3 (March 1979), pp. 315–26, http://onlinelibrary.wiley.com/doi/10.1111/j.1469-7998.1979.tb03372.x/full

Sakai, Sharon, Bradley M. Arsznov, Barbara Lundrigan and Kay E. Holekamp, "Brain Size and Social Complexity: A Computed

Tomography Study in Hyaenidae," *Brain, Behavior and Evolution* 77.2 (2011), pp. 91–104, https://www.ncbi.nlm.nih.gov/pubmed/21335942

Sax, Boria, *The Mythical Zoo: Animals in Life, Legend and Literature* (The Overlook Press, 2013)

Smith, Jennifer E., Joseph M. Kolowski, Katharine E. Graham, Stephanie E. Dawes and Kay E. Holekamp, "Social and Ecological Determinants of Fission–Fusion Dynamics in the Spotted Hyaena," *Animal Behaviour* 76.3 (September 2008), pp. 619–36, https://doi.org/10.1016/j.anbehav.2008.05.001

Szykman, Micaela, Russell C. Van Horn, Anne L. Engh, Erin E. Boydston and Kay E. Holekamp, "Courtship and Mating in Free-Living Spotted Hyenas," *Behaviour* 144.7 (July 2007), pp. 815–46, http://www.jstor.org/stable/4536481

Watson, Morrison, "On the Female Generative Organs of Hyaena *Crocuta*," *Proceedings of the Zoological Society of London* 24 (1877), pp. 369–79

Zimmer, Carl, "Sociable and Smart," *New York Times*, 4 March 2008

CHAPTER 5. VULTURE

Audubon, John James, "An Account of the Habits of the Turkey Buzzard (*Vultur aura*) Particularly with the View of Exploding the Opinion Generally Entertained of Its Extraordinary Power of Smelling," *Edinburgh New Philosophical Journal* 2 (Edinburgh: Adam Black, 1826)

Beck, Herbert H., "'The Occult Senses in Birds," *The Auk* 37 (1920), pp. 55–9

Birkhead, Tim, *Bird Sense: What It's Like to Be a Bird* (London: Bloomsbury, 2012)

Blackburn, Julia, *Charles Waterton, 1782–1865: Traveller and Conservationist* (London: Vintage, 1989)

Buffon, Georges-Louis Leclerc, Comte de, *The Natural History of Quadrupeds by the Count of Buffon; Translated from the French. With an Account of the Life of the Author* (Edinburgh: Thomas Nelson and Peter Brown, 1830).

Darlington, P. J., "Notes on the Senses of Vultures," *The Auk* 47.2 (1930), pp. 251–2

Dooren, Thom van, *Vulture* (London: Reaktion, 2011)

Dooren, Thom van, "Vultures and Their People in India: Equity and Entanglement in a Time of Extinctions," *Australian Humanities Review* 50 (May 2011), pp. 130–46, http://www.australianhumanitiesreview.org/archive/Issue-May-2011/vandooren.html

Gurney, J. H., "On the Sense of Smell Possessed by Birds," *Ibis* 4.2 (April 1922)

Henderson, Carrol L., *Birds in Flight: The Art and Science of How Birds Fly* (Minneapolis: Voyageur Press, 2008)

Houston, David C., "Scavenging Efficiency of Turkey Vultures in Tropical Forest," *Condor* 88.3 (1986), pp. 318–23, https://sora.unm.edu/sites/default/files/journals/condor/v088n03/p0318-p0323.pdf

Jackson, Andrew L., Graeme D. Ruxton and David C. Houston, "The Effect of Social Facilitation on Foraging Success in Vultures: A Modelling," *Biology Letters* 4.3 (23 June 2008), p. 311, http://rsbl.royalsocietypublishing.org/content/4/3/311

Kendall, Corinne J., Munir Z. Virani, J. Grant C. Hopcraft, Keith L. Bildstein and Daniel I. Rubenstein, "African Vultures Don't Follow Migratory Herds: Scavenger Habitat Use Is Not Mediated by Prey Abundance," *PLoS One* 9.1 (8 January 2014), https://doi.org/10.1371/journal.pone .0083470

Markandya, Anil, Tim Taylor, Alberto Longo, M. N. Murty, Sucheta Murty and Kishore Kumar Dhavala, "Counting the Cost of Vulture Decline: An Appraisal of the Human Health and Other Benefits of Vultures in India," *Ecological Economics* 67.2 (September 2008), pp. 194–204, http://dx.doi.org/10.1016/j.ecolecon.2008.04.020

Martin, Graham R., Steven J. Portugal and Campbell P. Murn, "Visual Fields, Foraging and Collision Vulnerability in *Gyps* Vultures," *Ibis* 154.3 (July 2012), pp. 626–31, http://onlinelibrary.wiley.com/doi/10.1111/j.1474919X.2012.01227.x/abstract

Rabenold, Patricia Parker, "Recruitment to Food in Black Vultures: Evidence for Following from Communal Roosts," *Animal Behaviour* 35.6 (December 1987), pp. 1775–85, http://www.sciencedirect.com/science/article/pii/S0003347287800702

Smith, Steven A., and Richard A. Paselk, "Olfactory Sensitivity of the Turkey Vulture (*Cathartes aura*) to Three Carrion-Associated Odorants," *The Auk* 103.3 (July 1986), pp. 586–92, http:///mambobob-raptorsnest .blogspot.co.uk/2008/02/olfactory-capabilities-in-t-rex-and.html

Stager, Kenneth E., "The Role of Olfaction in Food Location by the Turkey Vulture (*Cathartes aura*)," PhD thesis, University of Southern California (2014), https://nhm.org/site/sites/default/files/pdf/contrib_science/CS81.pdf

"Vultures," Vulture Conservation Foundation website, http://www.4 vultures.org/vultures

Waddell, Gene (ed.), *John Bachman: Selected Writings on Science, Race, and Religion* (Athens: University of Georgia Press, 2011)

Ward, Jennifer, Dominic J. McCafferty, David C. Houston and Graeme D. Ruxton, "Why Do Vultures have Bald Heads? The Role of Postural Adjustment and Bare Skin Areas in Thermoregulation," *Journal of Thermal Biology* 33.3 (April 2008), pp. 168–73, https://www.researchgate.net/publication/223457788

Waterton, Charles, *Essays on Natural History* (London: Frederick Warne, 1871)

Wilkinson, Benjamin Joel (dir.), *Carrion Dreams 2.0: A Chronicle of the Human-Vulture Relationship* (Abominationalist Productions, 2012)

CHAPTER 6. BAT

Allen, Glover M., *Bats: Biology, Behavior, and Folklore* (Mineola, NY: Dover Publications, 2004)

Boyles, Justin G., Paul M. Cryan, Gary F. McCracken and Thomas I I. Kunz, "Economic Importance of Bats in Agriculture," *Science* 332.6025 (1 April 2011), pp. 41–2, http://science.sciencemag.org/content/332/6025/41

Carter, Gerald G., and Gerald S. Wilkinson, "Food Sharing in Vampire Bats: Reciprocal Help Predicts Donations More than Relatedness or Harassment," *Proceedings of the Royal Society B* 280.1753 (22 February 2013), pp. 1–6, https://www.ncbi.nlm.nih.gov/pmc/articles/PMC35/4350

Chivers, Charlotte, "Why Isn't Everyone 'Batty' About Bats?," One Poll, 19 May 2015, http://www.onepoll.com/why-isnt-everyone-batty-about-bats

Dijkgraaf, Sven, "Spallanzani's Unpublished Experiments on the Sensory Basis of Object Perception in Bats," *Isis* 51.1 (1960), pp. 9–20

Ditmars, Raymond, "The Vampire Bat: A Presentation of Undescribed Habits and Review of its History," *Zoologica*, vol. XIX, no.2, 1935

Dodd, Kevin, *Blood Suckers Most Cruel: The Vampire and the Bat in and before Dracula* (Kevin Dodd, Visiting Scholar, Vanderbilt University)

Galambos, Robert, "The Avoidance of Obstacles by Flying Bats: Spallanzani's Ideas (1794) and Later Theories," *Isis* 34.2 (1942), pp. 132–40

Greenhall, Arthur, *Natural History of Vampire Bats* (CRC Press, 1988) Griffin, Donald R., *Listening in the Dark: The Acoustic Orientation of Bats and Men* (New Haven, CT: Yale University Press, 1958)

Gröger, Udo, and Lutz Wiegrebe, "Classification of Human Breathing Sounds by the Common Vampire Bat, *Desmodus rotundus*," *BMC Biology*

4.1 (16 June 2006), https://bmcbiol.biomedcentral.com/articles/10.1186
/1741-7007-4-18

McCracken, Gary F., "Bats and Vampires," *Bat Conservation International*
11.3 (Fall 1993), http://www.batcon.org/resources/media-education/
bats-magazine/bat_article/603

McCracken, Gary F., "Bats in Belfries and Other Places," *Bat Conservation
International* 10.4 (Winter 1992), http://www.batcon.org/resources/
media-education/bats-magazine

McCracken, Gary F. "Bats in Magic, Potions, and Medicinal Preparation,"
Bat Conservation International 10.3 (Fall 1992), http://www.batcon.org/
resources/media-education/bats-magazine/bat_article/546

Müller, Briggite, Martin Glösmann, Leo Peichl, Gabriel C. Knop, Cornelia
Hagemann and Josef Ammermüller, "Bat Eyes Have Ultraviolet-Sensitive
Cone Photoreceptors," *PLoS One* 4.7 (28 July 2009), p. e6390, https://doi
.org/10.1371/journal.pone.0006390

Pitnick, Scott, Kate E. Jones and Gerald S. Wilkinson, "Mating System and
Brain Size in Bats," *Proceedings of the Royal Society of London B* 273.1587 (22
March 2006), pp. 719–24

Riskin, Daniel K., and John W. Hermanson, "Biomechanics: Independent
Evolution of Running in Vampire Bats," *Nature* 434 (17 March 2005),
p. 292, https://www.nature.com/nature/journal/v434/n7031/full/
434292a.html

Schutt, Bill, *Dark Banquet: Blood and the Curious Lives of Blood-Feeding Creatures* (New York: Broadway Books, 2009)

Schutt, William A., J. Scott Altenbach, Young Hui Chang, Dennis M. Culli-
nane, John W. Hermanson, Farouk Muradali and John E. A. Bertram,
"The Dynamics of Flight-Initiating Jumps in the Common Vampire
Bat *Desmodus rotundus*," *Journal of Experimental Biology* 200.23 (1997), pp.
3003–12, http://jeb.biologists.org/content/200/23/3003

Surlykke, Annemarie, and Elisabeth K. V. Kalko, "Echolocating Bats Cry
Out Loud to Detect Their Prey," *PLoS One* 3.4 (30 April 2008), https://doi.
org/10.1371/journal.pone.0002036

Tan, Min, Gareth Jones, Guangjian Zhu, Jianping Ye, Tiyu Hong, Shanyi
Zhou, Shuyi Zhang and Libiao Zhang, "Fellatio by Fruit Bats Prolongs
Copulation Time," *PLoS One*, 4.10 (28 October 2009), https://doi.org/
10.1371/journal.pone.0007595

Wilkinson, Gerald S., "Social Grooming in the Common Vampire Bat, *Desmodus rotundus*," *Animal Behaviour* 34.6 (1986), pp. 1880–89

Wilson, E. O., and Stephen R. Kellert (eds.), *The Biophilia Hypothesis* (Washington, DC: Island Press, 1993)

CHAPTER 7. FROG

Berger, Lee, Richard Speare, Peter Daszak, D. Earl Green, Andrew A.Cunningham, C. Louise Goggin, Ron Slocombe, Mark A. Ragan, Alex D. Hyatt, Keith R. McDonald, Harry B. Hines, Karen R. Lips, Gerry Marantelli and Helen Parkes, "Chytridiomycosis Causes Amphibian Mortality Associated with Population Declines in the Rain Forests of Australia and Central America," *Proceedings of the National Academy of Sciences USA* 95.15 (21 July 1998), pp. 9031–6, http://www.pnas.org/content/95/15/9031.full

Bondeson, Jan, *The Feejee Mermaid: And Other Essays in Natural and Unnatural History* (Ithaca, NY: Cornell University Press, 1999)

Cobb, Matthew, *The Egg and Sperm Race: The Seventeenth-Century Scientists Who Unravelled the Secrets of Sex, Life, and Growth* (London: Simon & Schuster, 2007)

Collins, James P., Martha L. Crump and Thomas E. Lovejoy III, *Extinction in Our Times: Global Amphibian Decline* (Oxford: Oxford University Press, 2009)

Cousteau, Jacques (dir.), "Legend of Lake Titicaca," *The Undersea World of Jacques Cousteau* (Metromedia Productions, 1969)

Daston, Lorraine, and Elizabeth Lunbeck, *Histories of Scientific Observation* (Chicago: University of Chicago Press, 2011)

Gurdon, John B., and Nick Hopwood, "The Introduction of *Xenopus Laevis* into Developmental Biology: Of Empire, Pregnancy Testing and Ribosomal Genes," *International Journal of Developmental Biology* 44.1 (2003), pp. 43–50, http://www.ijdb.ehu.es/web/paper.php?doi=10761846

Hogben, Lancelot Thomas, *Lancelot Hogben, Scientific Humanist: An Unauthorised Autobiography* (London: Merlin Press, 1998)

Lips, Karen R., Forrest Brem, Roberto Brenes, John D. Reeve, Ross A. Alford, Jamie Voyles, Cynthia Carey, Lauren Livo, Allan P. Pessier and James P. Collins, "Emerging Infectious Disease and the Loss of Biodiversity in a Neotropical Amphibian Community," *Proceedings of the National Academy of Sciences USA* 103.9 (28 February 2006), pp. 3165–70, http://www.pnas.org/content/103/9/3165

McCartney, Eugene S., "Spontaneous Generation and Kindred Notions in Antiquity," *Transactions and Proceedings of the American Philological Association* 51 (1920), pp. 101–15, http://www.jstor.org/stable/282874

Olszynko-Gryn, Jesse, "Pregnancy Testing in Britain, c. 1900–67: Laboratories, Animals and Demand from Doctors, Patients and Consumers," PhD thesis, University of Cambridge (2015)

Oxford, Pete, and Renée Bish, "In the Land of Giant Frogs: Scientists Strive to Keep the World's Largest Aquatic Frog Off a Growing Global List of Fleeting Amphibians," 1 October 2003, https://www.nwf.org/News-and-Magazines/National-Wildlife/Animals/Archives/2003/In-the-Land-of-Giant-Frogs.aspx

Piper, Ross and Mike Shanahan, *Extraordinary Animals: An Encyclopedia of Curious and Unusual Animals* (Westport, CT: Greenwood, 2007)

Redi, Francesco, *Experiments on the Generation of Insects* (Chicago: Open Court Publishing Company, 1909)

Skerratt, Lee Francis, Lee Berger, Richard Speare, Scott Cashins, Keith R. McDonald, Andrea D. Phillott, Harry B. Hines and Nicole Kenyon, "Spread of Chytridiomycosis Has Caused the Rapid Global Decline and Extinction of Frogs," *EcoHealth* 4 (2007), pp. 125–34, https://link.springer.com/article/10.1007%2Fs10393-007-0093-5

Sleigh, Charlotte, *Frog* (London: Reaktion, 2012)

Soto-Azat, Claudio, Barry T. Clarke, John C. Poynton, Matthew Charles Fisher, S. F. Walker and Andrew A. Cunningham, "Non-Invasive Sampling Methods for the Detection of *Batrachochytrium dendrobatidis* in Archived Amphibians," *Diseases of Aquatic Organisms* 84.2 (6 April 2009), pp. 163–6, https://www.ncbi.nlm.nih.gov/pubmed/19476287

Soto-Azat, Claudio, Andés Valenzuela Sánchez, Ben Collen, J. Marcus Rowcliffe, Alberto Veloso and Andrew A. Cunningham, "The Population Decline and Extinction of Darwin's Frogs," *PLoS One* 8.6 (12 June 2013), p. e66957, https://www.ncbi.nlm.nih.gov/pmc/articles/PMC3680453

Soto-Azat, Claudio, Alexandra Peñafiel-Ricaurte, Stephen J. Price, Nicole Sallaberry-Pincheira, María Pía García, Mario Alvarado-Rybak and Andrew A. Cunningham, "*Xenopus laevis* and Emerging Ampibian Pathogens in Chile," *EcoHealth* 13.4 (December 2016), pp. 775–83, https://link.springer.com/article/10.1007/s10393-016-1186-9

Terrall, Mary, "Frogs on the Mantelpiece: The Practice of Observation in Daily Life," in Lorraine Daston and Elizabeth Lunbeck (eds.), *Histories of Scientific Observation* (Chicago: University of Chicago Press, 2011)

van Sittert, Lance, and G. John Measey, "Historical Perspectives on Global

Exports and Research of African Clawed Frogs (*Xenopus laevis*)," *Transactions of the Royal Society of South Africa* 71.2 (2016), pp. 157–66, http://www.tandfonline.com/doi/abs/10.1080/0035919X.2016.1158747.

Waller, John, *Leaps in the Dark: The Making of Scientific Reputations* (Oxford: Oxford University Press, 2004)

CHAPTER 8. STORK

Aldersey-Williams, Hugh, *The Adventures of Sir Thomas Browne in the Twenty-First Century* (London: Granta, 2015)

Aristotle, *History of Animals in Ten Books*, vols. 8–9, trans. by Richard Cresswell (London: George Bell, 1878)

Arnott, Geoffrey, *Birds in the Ancient World from A to Z* (Routledge, 2012)

Barrington, Daines, Miscellanies (London: Nichols, 1781)

Beattie, James, et al., *Eco-Cultural Networks of the British Empire* (Bloomsbury, 2014)

Birkhead, Tim, *Bird Sense: What It's Like to Be a Bird* (London: Bloomsbury, 2011)

Birkhead, Tim, *The Wisdom of Birds: An Illustrated History of Ornithology* (London: Bloomsbury, 2008)

Birkhead, Tim, Jo Wimpenny and Bob Montgomerie, *Ten Thousand Birds: Ornithology Since Darwin* (Princeton, NJ: Princeton University Press, 2014)

Bont, Raf de, *Stations in the Field: A History of Place-Based Animal Research, 1870–1930* (Chicago: University of Chicago Press, 2015)

Buffon, Georges-Louis Leclerc, Comte de, *The Book of Birds: Edited and Abridged from the Text of Buffon* (London: R. Tyas, 1841)

Cocker, Mark, and David Tipling, *Birds and People* (London: Jonathan Cape, 2013)

Cuvier, Georges, *The Animal Kingdom*, ed. by H. M'Murtrie (New York: Carvill, 1831)

Gerald of Wales, Topographia Hibernica, quoted in Patrick Armstrong, *The English Parson-Naturalist: A Companionship Between Science and Religion* (Leominster: Gracewing Publishing, 2000)

"Guide to North American Birds: Common Poorwill (Phalaenoptilus nuttallii)," National Audubon Society, http://www.audubon.org/field-guide/bird/common-poorwill

Harrison, C. J. O., "Pleistocene and Prehistoric Birds of South-west Britain," *Proceedings of the University of Bristol Spelaeological Society* 18.1 (1987), pp.

81–104, http://www.ubss.org.uk/resources/proceedings/vol18/UBSS_
Proc_18_1_81-104.pdf

Haverschmidt, F., *The Life of the White Stork* (Leiden: Brill Archive, 1949)

Kinzelbach, Ragnar K., Das Buch Vom Pfeilstorch (Berlin: Basilisken-Presse, 2005)

Lewis, Andrew J., *A Democracy of Facts: Natural History in the Early Republic*
(Philadelphia: University of Pennsylvania Press, 2011)

McCarthy, Michael J., *Say Goodbye to the Cuckoo* (London: John Murray, 2010)

McNamee, Gregory, *Aelian's on the Nature of Animals* (Dublin: Trinity University Press, 2011)

Park, Thomas (ed.), *The Harleian Miscellany: A Collection of Scarce, Curious, and
Entertaining Pamphlets and Tracts*, vol. 5 (London: White and Murray, 1810)

Rennie, James, *Natural History of Birds: Their Architecture, Habits, and Facul-
ties* (London: Harper, 1859)

Rickard, Bob, and John Michell, *The Rough Guide to Unexplained Phenomena*
(London: Penguin, 2010)

Simon, Matt, "Fantastically Wrong: The Scientist Who Thought That Birds
Migrate to the Moon," *Wired*, 22 October 2014, https://www.wired.com
/2014/10/fantastically-wrong-scientist-thought-birds-migrate-moon

Tate, Peter, *Flights of Fancy: Birds in Myth, Legend and Superstitio* (London:
Random House, 2007)

Turner, Angela, *Swallow* (London: Reaktion, 1994)

Vaughan, Richard, *Wings and Rings: A History of Bird Migration Studies in
Europe* (Penryn: Isabelline Books, 2009)

Wilcove, David S., and Martin Wikelski, "Going, Going, Gone: Is Animal
Migration Disappearing," *PLoS Biology* 6.7 (29 July 2008), http://journals
.plos.org/plosbiology/article?id=10.1371/journal.pbio.0060188

Wilkins, John, *The Discovery of a World in the Moone* (London: Sparke and
Forrest, 1638)

Witsen, Nicholaas, Emily O'Gorman and Edward Mellilo (eds.), *Beattie's
Eco-Cultural Networks and the British Empire: New Views on Environmental
History* (London: Bloomsbury, 2016)

CHAPTER 9. HIPPOPOTAMUS

Barklow, William E., "Amphibious Communication with Sound in Hippos,
Hippopotamus amphibius," *Animal Behaviour* 68.5 (2004), pp. 1125–32,
doi:10.1016/j.anbehav.2003.10.034

Bostock, John, and Henry T. Riley (eds.), *The Natural History of Pliny* (London: Henry G. Bohn, 1855)

Dawkins, Richard, *The Ancestor's Tale: A Pilgrimage to the Dawn of Life* (London: Weidenfeld & Nicolson, 2010)

Gatesy, John, "More DNA Support for a Cetacea/Hippopotamidae Clade: The Blood-Clotting Protein Gene Gamma-Fibrinogen," *Molecular Biology and Evolution* 14.5 (May 1997), pp. 537– 43, https://www. ncbi.nlm.nih.gov /pubmed/9159931

Grice, Gordon, *Book of Deadly Animals* (London: Penguin, 2012)

Kremer, William, "Pablo Escobar's Hippos: A Growing Problem," BBC News, 26 June 2014, http://www.bbc.com/news/magazine-27905743

Lihoreau, Fabrice, Jean-Renaud Boisserie, Frederick Kyalo Manthi and Stéphane Ducrocq, "Hippos Stem from the Longest Sequence of Terrestrial Cetartiodactyl Evolution in Africa," *Nature Communications* 6.6264 (24 February 2015), https://www.nature.com/articles/ncomms7264

Saikawa, Yoko, Kimiko Hashimoto, Masaya Nakata, Masato Yoshihara, Kiyoshi Nagai, Motoyasu Ida and Teruyuki Komiya, "Pigment Chemistry: The Red Sweat of the Hippopotamus," *Nature* 429 (27 May 2004), p. 363, https://www.nature.com/nature/journal/v429/n6990/full/429363a.html

Sax, Boria, *The Mythical Zoo: An Encyclopedia of Animals in World Myth, Legend, and Literature* (Santa Barbara, CA: ABC-Clio, 2001)

Thewissen, J. G. M. "Hans," *The Walking Whales: From Land to Water in Eight Million Years* (Berkeley: University of California Press, 2014)

Thompson, Ken, *Where Do Camels Belong?: The Story and Science of Invasive Species* (London: Profile, 2014)

CHAPTER 10. MOOSE

Ceaser, James W., *Reconstructing America: The Symbol of America in Modern Thought* (London: Yale University Press, 2000)

Dudley, Theodore Robert, *The Drunken Monkey: Why We Drink and Abuse Alcohol* (Berkeley: University of California Press, 2014)

Dugatkin, Lee Alan, *Mr. Jefferson and the Giant Moose: Natural History in Early America* (Chicago: University of Chicago Press, 2009)

Ford, Paul (ed.), *The Works of Thomas Jefferson: Correspondence and Papers, 1816–1826*, vol. 7 (New York: Cosimo Books, 2009)

Griggs, Walter S., and Frances P. Griggs, *A Moose's History of North America* (Richmond, VA: Brandylane Publishers, 2009)

Jackson, Kevin, *Moose* (London: Reaktion, 2008)

Jefferson, Thomas, *Notes on the State of Virginia* (Boston, MA: H. Sprague, 1802)

Merrill, Samuel, *The Moose Book: Facts and Stories from Northern Forests* (New York: Dutton, 1920)

Mooallem, Jon, *Wild Ones: A Sometimes Dismaying, Weirdly Reassuring Story About Looking at People Looking at Animals in America* (London: Penguin Books, 2014)

Morris, Steve, David Humphreys and Dan Reynolds, "Myth, Marula, and Elephant: An Assessment of Voluntary Ethanol Intoxication of the African Elephant (*Loxodonta africana*) Following Feeding on the Fruit of the Marula Tree (*Sclerocarya birrea*)," *Physiological and Biochemical Zoology* 79.2 (March/April 2006), pp. 363–9, http://www.journals.uchicago.edu/doi/abs/10.1086/499983

Mosley, Adam, *Bearing the Heavens: Tycho Brahe and the Astronomical Community of the Late Sixteenth Century* (Cambridge: Cambridge University Press, 2007)

Siegel, Ronald K., *Intoxication: The Universal Drive for Mind-Altering Substances* (Park Street Press, 1989)

Siegel, Ronald K., and Mark Brodie, "Alcohol Self-Administration by Elephants," *Bulletin of the Psychonomic Society* 22.1 (July 1984), https://link.springer.com/article/10.3758/BF03333758

CHAPTER II. PANDA

Becker, Elizabeth, *Overbooked: The Exploding Business of Travel and Tourism* (New York: Simon & Schuster, 2016)

Buckingham, Kathleen C., Jonathan Neil, William David and Paul R. Jepson, "Diplomats and Refugees: Panda Diplomacy, Soft 'Cuddly' Power, and the New Trajectory in Panda Conservation," *Environmental Practice* 15.3 (2013), pp. 262–70, https://www.researchgate.net/publication/255981642.

Christiansen, Per, and Stephen Wroe, "Bite Forces and Evolutionary Adaptations to Feeding Ecology in Carnivores," *Ecology* 88.2 (February 2007), pp. 347–58, https://www.jstor.org/stable/27651108

Conniff, Richard, *The Species Seekers: Heroes, Fools, and the Mad Pursuit of Life on Earth* (New York: W. W. Norton, 2010)

Cooke, Lucy, "The Power of Cute," BBC Radio4, http://www.bbc.co.uk/programmes/p03w3sxn

Croke, Vicky, *The Lady and the Panda: The True Adventures of the First American Explorer to Bring Back China's Most Exotic Animal* (New York: Random House, 2006)

Davis, D. Dwight, *The Giant Panda: A Morphological Study of Evolutionary Mechanisms* (Chicago: Natural History Museum, 1964)

Ellis, Susie, Anju Zhang, Hemin Zhang, Jinguo Zhang, Zhihe Zhang, Mabel Lam, Mark Edwards, JoGayle Howard, Donald Janssen, Eric Miller and David Wildt, "Biomedical Survey of Captive Giant Pandas: A Catalyst for Conservation Partnerships in China," in Donald Lindburg and Karen Baragona (eds.), *Giant Pandas: Biology and Conservation* (Berkeley: University of California Press, 2004), pp. 250–63, http://www.jstor.org/stable/10.1525/j.ctt1ppskn

"Giant Panda Feeding on Carrion," BBC Natural History Unit, http://www.arkive.org/giant-panda/ailuropoda-melanoleuca/video-08b.html (accessed 7 July 2017)

Graham-Jones, Oliver, *Zoo Doctor* (New York: Fontana Books, 1973)

Hagey, Lee R. and Edith A. MacDonald, "Chemical Composition of Giant Panda Scent and Its Use in Communication," in Donald Lindburg and Karen Baragona (eds.), *Giant Pandas: Biology and Conservation* (Berkeley: University of California Press, 2004), pp. 121– 4.

Hartig, Falk, "Panda Diplomacy: The Cutest Part of China's Public Diplomacy," *Hague Journal of Diplomacy* 8.1 (2013), pp. 49–78, https://eprints.qut.edu.au/59568

Hull, Vanessa, Jindong Zhang, Shiqiang Zhou, Jinuyan Huang, Rengui Li, Dian Liu, Weihua Xu, Yan Huang, Zhiyun Ouyang, Hemin Zhang and Jianguo Liu, "Space Use by Endangered Giant Pandas," *Journal of Mammalogy* 96.1 (2015), pp. 230–36, https://doi.org/10.1093/jmammal/gyu031

Lindburg, Donald, and K. Baragona (eds.), *Giant Pandas: Biology and Conservation* (Berkeley: University of California Press, 2004)

Morris, Ramona, and Desmond Morris, *Men and Pandas* (London: Hutchinson, 1966)

Nicholls, Henry, *Lonesome George: The Life and Loves of a Conservation Icon* (New York: Palgrave, 2007)

Nicholls, Henry, *Way of the Panda: The Curious History of China's Political Animal* (London: Profile, 2011)

Ringmar, Erik, "Audience for a Giraffe: European Exceptionalism and the Quest for the Exotic," *Journal of World History* 17.4 (December 2006), pp. 375–97

Schaller, George, *The Last Panda* (Chicago: University of Chicago Press, 1994)

Schaller, George, Hu Jinchu, Pan Wenshi and Zhu Jing, *The Giant Pandas of Wolong* (Chicago: University of Chicago Press, 1985)

White, Angela M., Ronald R. Swaisgood, Hemin Zhang, "The Highs and Lows of Chemical Communication in Giant Pandas (Ailuropoda melanoleuca): Effect of Scent Deposition Height on Signal Discrimination," *Behavioural Ecology Sociobiology* 51.6 (May 2002), pp. 519–29

Zhang, Peixun, Tianbing Wang, Jian Xiong, Feng Xue, Hailin Xu, Jian-hai Chen, Dianying Zhang, Zhongguo Fu and Baoguo Jiang, "Three Cases Giant Panda Attack on Human at Beijing Zoo," *International Journal of Clinical and Experimental Medicine* 7.11 (2014), pp. 4515–18, https://www.ncbi.nlm.nih.gov/pmc/articles/PMC4276236

Zhao, Shancen, Pingping Zheng, Shanshan Dong, Xiangjiang Zhan, Qi Wu, Xiaosen Guo, Yibo Hu, Weiming He, Shanning Zhang, Wei Fan, Lifeng Zhu, Dong Li, Xuemei Zhang, Quan Chen, Hemin Zhang, Zhihe Zhang, Xuelin Jin, Jinguo Zhang, Huanming Yang, Jian Wang, Jun Wang and Fuwen Wei, "Whole-Genome Sequencing of Giant Pandas Provides Insights into Demographic History and Local Adaptation," *Nature Genetics* 45.1 (January 2013), pp. 67–71, http://www.nature.com/ng/journal/v45/n1/full/ng.2494.html

CHAPTER 12. PENGUIN

Bagemihl, Bruce, *Biological Exuberance: Animal Homosexuality and Natural Diversity* (New York: St Martin's Press, 1999)

Bried, Joël, Frédéric Jiguet and Pierre Jouventin, "Why Do *Aptenodytes* Penguins Have High Divorce Rates?," *The Auk* 116.2 (1999), pp. 504–12, https://sora.unm.edu/sites/default/files/journals/auk/v116n02/p0504-p0512.pdf

Cherry-Garrard, Apsley, *The Worst Journey in the World: Antarctic, 1910–1913*, vol. 2 (New York: George H. Doran, 1922)

Clayton, William, "An Account of Falkland Islands," *Philosophical Transactions of the Royal Society of London* 66 (1 January 1776), pp. 99–108, http://rstl.royalsocietypublishing.org/content/66/99.full.pdf+html

Davis, Lloyd S., and Martin Renner, *The Penguins* (London: Bloomsbury, 2010)

Davis, Lloyd S., Fiona M. Hunter, Robert G. Harcourt and Sue Michelsen Heath, "Short Communication: Reciprocal Homosexual Mounting, Adélie Penguins *Pygoscelis adeliae*," *Emu* 98.2 (2001), pp. 136–7, http://www.publish.csiro.au/mu/MU98015

Fuller, Errol, *The Great Auk: The Extinction of the Original Penguin* (Piermont, NH: Bunker Hill Publishing, 2003)

Gurney, Alan, *Below the Convergence: Voyages Toward Antarctica, 1699–1839* (New York: W. W. Norton, 2007)

Haeckel, Ernst, *The Riddle of the Universe at the Close of the Nineteenth Century* (New York: Harper, 1905)

Hunter, Fiona M., and Lloyd S. Davis, "Female Adélie Penguins Acquire Nest Material from Extrapair Males After Engaging, Extrapair Copulations," *The Auk* 115.2 (April 1998), pp. 526–8, http://www.jstor.org/stable/4089218

Jacquet, Luc, and Bonne Pioche (dirs), *March of the Penguins* (National Geographic Films, 2005)

Larson, E. J., *An Empire of Ice: Scott, Shackleton, and the Heroic Age of Antarctic Science* (London: Yale University Press, 2011)

Martin, Stephen, *Penguin* (London: Reaktion, 2009)

Narborough, John, Abel Tasman, John Wood and Friderich Martens, *An Account of Several Late Voyages and Discoveries to the South and North* (Cambridge: Cambridge University Press, 2014; f.p. 1711)

Roy, Tui de, Mark Jones and Julie Cornthwaite, *Penguins: The Ultimate Guide* (Princeton, NJ: Princeton University Press, 2014)

Russell, Douglas G. D., William J. I. Sladen and David G. Ainley, "Dr. George Murray Levick (1876–1956): Unpublished Notes on the Sexual Habits of the Adélie Penguin," *Polar Record* 48.4 (October 2012), pp. 387–93, https://doi.org/10.1017/S0032247412000216

Wheeler, Sara, *Cherry: A Life of Apsley Cherry-Garrard* (London: Vintage, 2007)

Williams, T. D., "Mate Fidelity, Penguins," *Oxford Ornithology Series* 6.1, pp. 268–85

Wilson, Edward A., *Report on the Mammals and Birds, National Antarctic Expedition 1901–1904*, vol. 2 (London: Aves, 1907)

Wilson, Edward A., and T. G. Taylor, *With Scott: The Silver Lining* (New York: Dodd, Mead and Company, 1916)

CHAPTER 13. CHIMPANZEE

Bedford, J. M., "Sperm/Egg Interaction: The Specificity of Human Spermatozoa," *Anatomical Record*, 188 (1977), pp. 477–87. doi:10.1002/ar.1091880407

Buffon, Georges-Louis Leclerc, Comte de, *History of Quadrupeds*, vol. 3 (Edinburgh: Thomas Nelson, 1830)

Cohen, Jon, *Almost Chimpanzee: Redrawing the Lines that Separate Us from Them* (London: St Martin's Press, 2002)

Crockford, Catherine, Roman M. Wittig, Roger Mundry and Klaus Zuber-bühler, "Wild Chimpanzees Inform Ignorant Group Members of Danger," *Current Biology* 22.2 (24 January 2012), pp. 142–6, https://www.ncbi.nlm.nih.gov/pubmed/22209531

Cuperschmid, E. M. and T. P. R. D. Campos, "Dr. Voronoff's Curious Glandular Xeno-Implants," *História, Ciências, Saúde-Manguinhos* 14.3 (2007), pp. 737–60

de Waal, Frans, and Jennifer J. Pokorny, "Faces and Behinds: Chimpanzee Sex Perception," *Advanced Science Letters* 1.1 (June 2008), pp. 99–103, https://doi.org/10.1166/asl.2008.006

Gould, Stephen Jay, *Leonardo's Mountain of Clams and the Diet of Worms* (Cambridge, MA: Harvard University Press, 2011)

Gross, Charles, "Hippocampus Minor and Man's Place in Nature: A Case Study in the Social Construction of Neuroanatomy," *Hippocampus* 3.4 (1993), pp. 403–16

Hawks, John, "How Strong Is a Chimpanzee, Really?," *Slate*, http://www.slate.com/articles/health_and_science/science/2009/02/how_strong_is_a_chimpanzee.html

Hobaiter, Cat, and Richard W. Byrne, "The Meanings of Chimpanzee Gestures," *Current Biology* 24.14 (21 July 2014), pp. 1596–600, https://www.ncbi.nlm.nih.gov/pubmed/24998524

Hockings, Kimberley J., Nicola Bryson-Morrison, Susana Carvalho, Michiko Fujisawa, Tatyana Humle, William C. McGrew, Miho Nakamura, Gaku Ohashi, Yumi Yamanashi, Gen Yamakoshi and Tetsuro Matsuzawa, "Tools to Tipple: Ethanol Ingestion by Wild Chimpanzees Using Leaf-Sponges," *Royal Society: Open Science* 2.6 (9 June 2015), http://rsos.royalsocietypublishing.org/content/2/6/150150

IUCN, "Four Out of Six Great Apes One Step Away from Extinction—IUCN Red List," 2016, https://www.iucn.org/news/species/201609/four-out-six-great-apes-one-step-away-extinction-%E2%80%93-iucn-red-list (accessed 6 May 2017)

Janson, H. W., *Apes and Ape Lore in the Middle Ages and the Renaissance* (London: Warburg Institute, 1952)

Kahlenberg, Sonya M., and Richard W. Wrangham, "Sex Differences in Chimpanzees' Use of Sticks as Play Objects Resemble Those of Children," *Current Biology* 20.24 (21 December 2010), pp. R1067–8, http://dx.doi.org/10.1016/j.cub.2010.11.024

Kühl, Hjalmar S., Ammie S. Kalan, Mimi Arandjelovic, Floris Aubert, et al., "Chimpanzee Accumulative Stone Throwing," *Scientific Reports* 6 (29 February 2016), https://www.nature.com/articles/srep22219

Lucas, J. R., "Wilberforce and Huxley: A Legendary Encounter," *Historical Journal* 22.2 (1979)

Marks, Jonathan, *What It Means to Be 98% Chimpanzee: Apes, People, and Their Genes* (Berkeley: University of California Press, 2002)

Owen, Richard, "On the Characters, Principles of Division, and Primary Groups of the Class Mammalia," *Journal of the Proceedings of the Linnean Society I : Zoology* (London: Longman, 1857)

Pain, Stephanie, "Blasts from the Past: The Soviet Ape-Man Scandal," *New Scientist*, 2008, https://www.newscientist.com/article/mg19926701-000 -blasts-from-the-past-the-soviet-ape-man-scandal (accessed 5 May 2017)

Patterson, Nick, Daniel J. Richter, Sante Gnerre, Eric S. Lander and David Reich, "Genetic Evidence for Complex Speciation of Humans and Chimpanzees," *Nature* 441 (29 June 2006), pp. 1103–8, https://www.nature.com /nature/journal/v441/n7097/full/nature04789.html

Pliny the Elder, *The Natural History*, trans. by H. Rackham (London: William Heinemann, 1940)

Pruetz, Jill D., Paco Bertolani, Kelly Boyer Ontl, Stacy Lindshield, Mack Shelley and Erin G. Wessling, "New Evidence on the Tool-Assisted Hunting Exhibited by Chimpanzees (*Pan troglodytes verus*) in a Savannah Habitat at Fongoli, Sénégal," *Royal Society: Open Science* 2.4 (15 April 2015), http://rsos.royalsocietypublishing.org/content/2/4/140507

Rossiianov, Kirill, "Beyond Species: Il'ya Ivanov and His Experiments on Cross-Breeding Humans with Anthropoid Apes," *Science in Context* 15.2 (2002), pp. 277–316, https://www.cambridge.org/core/journals/science -in-context/article/div-classtitlebeyond-species-ilya-ivanov-and-his -experiments-on-cross-breeding-humans-with-anthropoid-apesdiv/ D3E0E117E953A0038D63984A D92F4B80

Sax, Boria, *The Mythical Zoo: An Encyclopedia of Animals in World Myth, Legend, and Literature* (Santa Barbara, CA: ABC-Clio, 2001)

Schwartz, Jeffrey H., *Orangutan Biology* (Oxford: Oxford University Press, 1988)

Sorenson, John, *Ape* (Reaktion, 2009)

Temerlin, Maurice K., *Lucy: Growing Up Human—A Chimpanzee Daughter in a Psychotherapist's Family* (Palo Alto, CA: Science & Behavior Books, 1975)

Topsell, Edward, *The History of Four-Footed Beasts and Serpents and Insects*, vol. 1 (New York: DaCapo, 1967; f.p. 1658)

Yerkes, Robert, and Ada Yerkes, *The Great Apes: A Study of Anthropoid Life* (New Haven, CT: Yale University Press, 1929)

Zimmer, Carl, "Searching for Your Inner Chimp," *Natural History*, December 2002–January 2003, http://www.carlzimmer.com/articles/PDF/02.ChimpDNA .pdf

CONCLUSION

de Waal, Frans, "Do Animals Have Morals?" http://www.npr.org/2014/08/15/338936897/do-animals-have-morals

Mills, Brett, "The Animals Went in Two by Two: Heteronormativity in Television Wildlife Documentaries," *European Journal of Cultural Studies* 16(1), pp. 100–114. ©The Author(s) 2012, reprints and permission: sagepub .co.uk/journalsPermissions.nav DOI: 10.1177/1367549412457477

Index

Lucy Cooke is an award-winning filmmaker who has written, produced, and directed several popular documentary series for the BBC, PBS, Discovery, and National Geographic. Her first book, *A Little Book of Sloth*, was a *New York Times* bestseller. She has written for the *Wall Street Journal* and the *New York Times*, among others, and she holds an MA in zoology from the University of Oxford. She lives in London.

Photo credit: David Dunkerley